Integrated Process Planning, Scheduling, and Due-Date Assignment

Traditionally, the three most important manufacturing functions are process planning, scheduling, and due-date assignment, which are handled sequentially and separately. This book integrates these manufacturing processes and functions to increase global performance along with manufacturing and production cost savings.

Integrated Process Planning, Scheduling, and Due-Date Assignment combines the most important manufacturing functions to use manufacturing resources better, reduce production costs, and eliminate bottlenecks with increased production efficiency. The book covers how the integration will help eliminate scheduling conflicts and how to adapt to irregular shop floor disturbances. It also explains how other elements, such as tardiness and earliness, are penalized and how prioritizing helps improve weight performance function.

This book will draw the interest of professionals, students, and academicians in process planning, scheduling, and due-date assignment. It could also be supplemental material for manufacturing courses in industrial engineering and manufacturing engineering departments.

Integrated Process Planning, Scheduling, and Due-Date Assignment

Halil İbrahim Demir
Abdullah Hulusi Kökçam
Caner Erden

CRC Press
Taylor & Francis Group
Boca Raton London New York

CRC Press is an imprint of the
Taylor & Francis Group, an **informa** business

Designed cover image: Shutterstock

First edition published 2024
by CRC Press
2385 Executive Center Drive, Suite 320, Boca Raton FL 33431

and by CRC Press
4 Park Square, Milton Park, Abingdon, Oxon, OX14 4RN

CRC Press is an imprint of Taylor & Francis Group, LLC

© 2024 Halil İbrahim Demir, Abdullah Hulusi Kökçam, Caner Erden

Library of Congress Cataloging-in-Publication Data
Names: Demir, Halil Ibrahim, author.
Title: Integrated process planning, scheduling, and due-date assignment /
Halil Ibrahim Demir, Abdullah Hulusi Kökçam, Caner Erden.
Description: First edition. | Boca Raton : CRC Press, [2024] | Includes
bibliographical references and index.
Identifiers: LCCN 2023005461 (print) | LCCN 2023005462 (ebook) | ISBN
9781032104263 (hbk) | ISBN 9781032104294 (pbk) | ISBN 9781003215295 (ebk)
Subjects: LCSH: Production control.
Classification: LCC TS157 .D445 2024 (print) | LCC TS157 (ebook) | DDC
658.5--dc23/eng/20230227
LC record available at https://lccn.loc.gov/2023005461
LC ebook record available at https://lccn.loc.gov/2023005462

ISBN: 978-1-032-10426-3 (hbk)
ISBN: 978-1-032-10429-4 (pbk)
ISBN: 978-1-003-21529-5 (ebk)

DOI: 10.1201/9781003215295

Typeset in Times
by SPi Technologies India Pvt Ltd (Straive)

Dedication

To my devoted mother and father (Elif and Gazi) and my precious wife and beloved daughter (Gülşen and Elif Özge).

Halil İbrahim Demir

To my invaluable parents (Makbule Ahsen and İbrahim) and to my precious family (Süheyla and Meryem), bestowed upon me by Allah the Glorified and the Exalted,

Abdullah Hulusi Kökçam

To my parents(Güler & Adalet), who have always been my inspiration and support.

To my wife, Sumeyye, who has been my rock and my partner in every step of the way.

To my daughter, Meryem İpek, my son, Ömer, who bring joy and laughter to my life every day. I hope they will grow up to be valuable citizens of our country, nation and the world.

Caner Erden

Contents

Preface

In today's ever-changing manufacturing environment, efficient and effective process planning, scheduling, and due date assignment functions are more important than ever. This book presents a comprehensive approach to integrated process planning, scheduling, and due date assignment. We aim to provide readers with a thorough understanding of this field's key concepts and techniques and practical guidance on applying them in real-world situations. As only a single problem has difficulty solving with deterministic methods with current hardware and software technologies due to the largeness of the solution space, the integrated problem is much more difficult. Thus, with current knowledge, metaheuristic methods are required to find near-optimal solutions. Therefore, an integration problem is presented, especially with metaheuristic algorithms.

This book is divided into twelve chapters. The first chapter introduces the concept of integrated process planning, scheduling, and due-date assignment, including key definitions and the importance of integrating manufacturing functions. Also, it discusses the problem's computational complexity and provides an overview of the book's structure and content. Chapters 2, 3, and 4 review the literature on single manufacturing functions and introduce classical approaches to those functions. They discuss the integration of the functions and their advantages over classical approaches.

Chapters 5 and 6 review the literature on integrated process planning and scheduling, introduce mathematical models and solutions and cover metaheuristic solutions and the challenges of integrated process planning and scheduling. Also, they discuss the challenges of dynamic integrated process planning and scheduling.

Chapters 7 and 8 present integration models for scheduling with due-date assignments and due-window assignments. Also, they review the literature on the topic covering mathematical and metaheuristic solutions for scheduling and due-date assignments.

Chapters 9 and 10 focus on the three functions' integration models and review the topic's literature. Also, they discuss metaheuristic solutions for integrated process planning and scheduling with due-date assignments in a dynamic environment.

Chapter 11 presents various solution techniques for integrated process planning, scheduling, and due-date assignment, such as exact, heuristic, metaheuristic, agent-based, and hybrid methods.

Chapter 12 presents the concept of integrating delivery as the fourth function and the importance of delivery in manufacturing. It covers the literature on delivery and vehicle routing and discusses scheduling with delivery and due-date assignment. Also, it concludes with the future of manufacturing.

The objective of this book can be summarized as follows:

- To promote preparing alternative process planning
- To integrate process planning with scheduling in the industry
- To provide better due dates when considered concurrently with process planning and scheduling

The book's intended audience includes, but is not limited to, students, academicians, and business people interested in process planning, scheduling, and due-date assignment subjects, especially those who wonder about how to integrate these important functions and simultaneously solve this complicated problem.

The book can be used as a reference for a separate course or be addressed in manufacturing courses for master's and doctoral students in industrial engineering and manufacturing engineering departments.

We hope this book will provide valuable insights and practical guidance to anyone looking for improving their production planning and scheduling capabilities.

Halil İbrahim Demir
Abdullah Hulusi Kökçam
Caner Erden
Sakarya, Türkiye
February 2023

Authors

Halil İbrahim Demir completed his undergraduate education at Bilkent University, Industrial Engineering Department, with a full scholarship in Ankara, Türkiye. He completed his master's degree in Industrial Engineering at Lehigh University, Bethlehem, USA, with a full scholarship from the Ministry of National Education of Türkiye. He started his PhD in Industrial Engineering at Northeastern University, Boston, USA, and completed his courses there. After returning to Türkiye, he finished his PhD thesis on "Integrated Process Planning, Scheduling, and Due-date Assignment" at Sakarya University. He is still working as Assistant Professor at Sakarya University. His research interests include optimization, simulation, scheduling, decision theory, multicriteria decision-making, and metaheuristic algorithms.

Abdullah Hulusi Kökçam completed his master's degree at Sakarya University, Department of Industrial Engineering, in 2010 with a study on fuzzy project scheduling using metaheuristic methods. He received his PhD from Sakarya University, Department of Industrial Engineering, in 2017. In his doctoral dissertation, he worked on an alternative model to evaluate school success in the entrance examination for higher education. He works as an assistant professor at the Sakarya University Industrial Engineering Department. His research interests include metaheuristic methods, artificial intelligence, fuzzy logic, optimization, and scheduling.

Caner Erden is currently assistant professor at the Sakarya University of Applied Sciences and a senior researcher at the AI Research and Application Center, Sakarya University of Applied Sciences, Sakarya, Türkiye. He worked as a research assistant in industrial engineering at Sakarya University and a researcher at Sakarya University AI Systems Application and Research between 2012 and 2020. He received his PhD from Sakarya University in the field of industrial engineering. His research interests include scheduling, discrete event simulation, metaheuristic algorithms, modeling and optimization, machine learning, and deep learning.

1 Introduction to Integrated Process Planning, Scheduling, and Due-Date Assignment

1.1 INTRODUCTION

With the increasing competition, it is essential to ensure that products are produced most effectively and efficiently and to deliver them to the customer on time to survive in the market. In this long process, there are many different and important steps, from the design of the product to be produced to the selection of the material to be used in the product, from the determination of the process operations to the amount of production, from when to produce which product and when to deliver it to the customer. These steps, which we call production functions, can be counted as materials, machines and equipment, methods, process planning (routing), estimating, loading and scheduling, due-date assignment, dispatching, expediting, inspection, and evaluation (Kashyap, 2015).

We will focus on the three major production functions: process planning, scheduling, and due-date assignment. From this point on, these three functions are introduced, and in later chapters, they will be given in detail. When we deal with the problems in these functions, their solutions take much time to solve as the problem size increases. This topic is investigated within the computational complexity area, which is given as a subtopic in this chapter. Although these functions are usually evaluated individually, they affect each other because they are bonded tightly, as they use one another's output as an input. Finally, we will discuss the importance of integrating these functions at the end of this chapter. More details with application examples have been given in other chapters.

1.2 PROCESS PLANNING

Process planning is an activity that links product design and manufacturing, combining manufacturing process knowledge with custom design under the limitation of workshop or factory manufacturing resources and preparing specific operation instructions (Li and Gao, 2020).

DOI: 10.1201/9781003215295-1

Process planning in production is used broadly, including manufacturing planning, material handling, process engineering, and the determination of machine routes. Each production stage is handled in detail with process planning, and the sequence of operations or processes required to manufacture or assemble a part is determined. Thus, it is ensured that the production is carried out more efficiently and that higher-quality outputs are obtained. As a result of process planning, the route to be followed in the production process and the operations to be processed are determined, which specify the order of all the processes required for the completion of the product. Process planning becomes much more important, especially in businesses where the same product is rarely made, or project-type production is carried out. Process planning is used in production and service sectors to determine how the service will be provided (Swamidass, 2000).

Process planning requires the utilization of many disciplines, including sequencing, machine selection, time and method study, programming, and material flow. Computer-aided process planning (CAPP) was developed in order to reduce the problems such as the lack of experienced personnel in process planning, which is traditionally carried out manually and experimentally, the low efficiency of the determined routes, the preparation of inconsistent routes by the personnel with different experiences, and the delay of the necessary reaction in the real production environment (Architecture Technology Corporation, 1991).

More detailed information about process planning is given in Chapter 2.

1.3 SCHEDULING

Scheduling is a common decision-making process in many manufacturing and service industries. It allocates resources to tasks in specific periods and aims to optimize one or more objectives. Scheduling, which has an important role in manufacturing and production systems, also has an important area of use in many service sectors, including transportation and distribution. In production, scheduling is often performed interactively with other decision support systems. Long-term plans determined by production planning affect scheduling. In addition, it may be necessary to update the schedule created in line with the information coming from the workshop (Pinedo, 2012).

Scheduling concerns questions such as deals with resource allocation, sequence of jobs, start and end time of jobs:

Which machine will do which job?
When will an operation/job start and end?
With what equipment and by whom will the work be done?
What will be the order of operations/jobs?

More detailed information about scheduling in manufacturing is given in Chapter 3.

1.4 DUE-DATE ASSIGNMENT

The due date is the time promised to the client that the job will be delivered. Each task has a specific priority level, earliest start time, and due date. Minimization of the number of jobs whose deadlines are delayed can be determined as a goal in scheduling (Pinedo, 2012). Many different due-date assignment rules affect customer and company relationships. It can be determined based on the number of tardy jobs or total tardiness. It is very crucial to give realistic due dates to the customers. If a company cannot keep its promise, its reputation suffers, and it may lose customers. A company may choose to give longer due dates to keep its promise, but at this time, it cannot compete in today's harsh competitive environment. Giving shorter due dates and trying to keep them may increase costs by keeping more buffer stock, paying overtime, etc. Both tardiness and earliness are not desired. With this in mind, a company can categorize its customers as significant or less significant and try to give shorter due dates to its significant customers by increasing its weight when deciding on due dates. Details of the due date assignment are given in Chapter 4.

1.5 COMPUTATIONAL COMPLEXITY

The computational complexity of a combinatorial problem depends on the computational behavior of the algorithms designed for its solution. Computational behavior is usually measured by the algorithm's running time, depending on the size of the problem (Florian et al., 1980). Some combinatorial optimization problems and their explanations are given as follows:

- Vehicle routing problem (VRP): VRP is an integer programming and combinatorial optimization problem that includes determining the best possible routes for a fleet of vehicles to take in order to deliver items to a specified group of clients. The entire route cost—monetary, physical, or other—is minimized. The VRP can save up to 30% on costs and has numerous industrial uses. Due to the NP-hardness of the problem, most solutions are found through heuristics (Toth and Vigo, 2002).
- Traveling salesman problem (TSP): Finding the shortest path between the points or places that must be visited is the goal of the algorithmic challenge of TSP. The cities a salesperson might visit are the points in the problem statement. The salesman wants to travel as little as possible in terms of distance and travel expenses. TSP, which focuses on optimization, is frequently used in computer science to determine the fastest path for the data provided to move between different nodes. Applications include finding hardware or network optimization techniques (Grötschel and Padberg, 1979).
- Minimum spanning tree (MST): One of the most well-known problems in graph theory is the MST problem. It entails determining the smallest number of edges, or the edges with the lowest weight, required to connect every vertex in a graph. Finding the least expensive route to connect nodes in a network, such as a computer or a phone network, is useful. It can also find a graph path with the lowest cost between two points. Some of the most well-liked algorithms for solving the MST issue are Kruskal's algorithm, Prim's algorithm, and Boruvka's algorithm (Graham and Hell, 1985).

- Linear programming (LP): A mathematical modeling technique called linear programming is used to choose the best result or solution from a set of constraints. It entails satisfying a set of restrictions while maximizing or minimizing a linear function. Complex optimization issues like resource allocation and scheduling can be resolved using it (Murota, 2020).
- Integer programming (IP): IP is a subset of the greater discipline of linear programming. Finding the best values when some or all of the variables are forced to have integer values (either in the minimization or maximization) is what this implies (Wolsey, 2020).
- Eight queens puzzle: A solution to the eight queens puzzle necessitates that no chess queens occupy the same row, column, or diagonal to ensure that no two queens threaten one another. The task is to arrange eight queens on a standard chess board without any of them assaulting any of the others. Placing eight queens on a chessboard without them protecting (or attacking) any of the others is the goal of the eight queens puzzle (Langendoen and Vree, 1992).
- Backpack problem: The backpack problem, also called the Knapsack Problem in computer science, is an optimization issue. The objective is to pack as much value as possible into a fixed-capacity backpack. It is an iconic illustration of a combinatorial optimization problem that aims to maximize the benefit of a group of objects while staying within the bounds of the knapsack. Branch and bound, greedy algorithms or dynamic programming can all be used to address the backpack problem (Talbott et al., 2009).
- Stock reduction problem: Stock reduction, sometimes referred to as inventory reduction, is the act of reducing the amount of stock carried by an organization to reduce expenses related to maintaining surplus inventory. It entails examining inventory levels, locating slow-moving or outmoded products, and putting plans to lower total stock levels. Some stock reduction tactics include shorter order cycles, decreased supplier lead times, and inventory optimization software (Cheng et al., 1994).
- Scheduling: The scheduling of a production line to maximize efficiency and reduce costs is known as the production scheduling problem. It entails choosing the ideal sequence to manufacture goods while considering the number and kind of goods, resource availability, and production process restrictions. The production scheduling issue can be solved via simulation, heuristics, or mathematical programming (Pinedo, 2012).

Process planning, scheduling, and due-date assignment problems encounter many uncertainties and constraints. Therefore, such problems are classified as NP (Non-deterministic Polynomial-time)—hard problems. Considering the computational complexity in solving only one function, the complexity encountered when these functions are considered together is multiplied. For this reason, obtaining the optimum solution takes a long time or is impossible with the current algorithms and the computing power of today's computers. At this point, metaheuristic algorithms have been developed that do not guarantee to reach the optimum solution but can obtain near-optimal solutions in a reasonable time.

1.6 IMPORTANCE OF INTEGRATING MANUFACTURING FUNCTIONS

Process planning, scheduling, and due-date assignment functions are inseparable parts of production. The scheduling function can be executed by using the product routes obtained as a result of the process planning and the process operations of the product as inputs. The due-date assignment function is realized according to the program that emerges from the scheduling. Therefore, each of these functions is run using the output from the other. Traditionally, these three functions were operated separately, but this led to the creation of disjointed plans and reduced efficiency. Afterward, a solution was achieved by integrating process planning and scheduling (IPPS). More detail about IPPS can be found in Chapter 5, and IPPS problems in a dynamic production environment (DIPPS) can be found in Chapter 6.

In addition, studies were carried out on integrating scheduling and due-date assignment functions (SWDDA) and due-window assignment functions (SWDWA). More details about SWDDA and SWDWA can be found in Chapters 7 and 8, respectively. Currently, studies are carried out on the integrated process planning, scheduling, and due-date assignment (IPPSDDA) problem by considering the evaluation of these three functions together. More detail about IPPSDDA can be found in Chapter 9. Furthermore, the IPPSDDA problem is also studied in a dynamic production environment (DIPPSDDA), which is also given in detail in Chapter 10. Solution techniques used in integrated manufacturing functions are discussed in Chapter 11.

In the future, it is inevitable to develop holistic approaches that will increase the scope of planning and enable it to be viewed from a broader perspective. Therefore integration of delivery as a fourth function of manufacturing and supply chain is discussed in detail in Chapter 12.

REFERENCES

Architecture Technology Corporation, 1991. *Computer Aided Process Planning (CAPP)*, 2nd ed. Elsevier Advanced Technology, Amsterdam and New York.

Cheng, C.H., Feiring, B.R., Cheng, T.C.E., 1994. The cutting stock problem—a survey. *Int. J. Prod. Econ.* 36, 291–305. https://doi.org/10.1016/0925-5273(94)00045-X

Florian, M., Lenstra, J.K., Kan, A.H.G.R., 1980. Deterministic production planning: algorithms and complexity. *Manag. Sci.* 26, 669–679.

Graham, R.L., Hell, P., 1985. On the history of the minimum spanning tree problem. *Ann. Hist. Comput.* 7, 43–57. https://doi.org/10.1109/MAHC.1985.10011

Grötschel, M., Padberg, M.W., 1979. On the symmetric travelling salesman problem I: inequalities. *Math. Program.* 16, 265–280.

Kashyap, D., 2015. *Production Planning and Control: (10 Functions)*. Your Artic. Libr. URL https://www.yourarticlelibrary.com/production-management/production-planning/production-planning-and-control-10-functions/57433 (accessed 9.1.21).

Langendoen, K.G., Vree, W.G., 1992. Eight queens divided: an experience in parallel functional programming, in: John Darlington, Roland Dietrich, ed. *Declarative Programming, Sasbachwalden 1991*. Springer, London, UK, pp. 101–115.

Li, X., Gao, L., 2020. *Effective Methods for Integrated Process Planning and Scheduling, Engineering Applications of Computational Methods*. Springer-Verlag, Berlin and Heidelberg. https://doi.org/10.1007/978-3-662-55305-3

Murota, K., 2020. Linear programming, in: Katsushi Ikeuchi, Koichiro Deguchi, eds. *Computer Vision: A Reference Guide*. Springer, pp. 1–7. Cham, Switzerland. https://doi.org/10.1007/978-3-030-03243-2_648-1

Pinedo, M.L., 2012. *Scheduling: Theory, Algorithms, and Systems*, 4th ed. Springer-Verlag, New York. https://doi.org/10.1007/978-1-4614-2361-4.

Swamidass, P.M. (Ed.), 2000. Process planning, in: *Encyclopedia of Production and Manufacturing Management*. Springer, Boston, MA, pp. 552–553. https://doi.org/10.1007/1-4020-0612-8_721

Talbott, N.R., Bhattacharya, A., Davis, K.G., Shukla, R., Levin, L., 2009. School backpacks: it's more than just a weight problem. *Work* 34, 481–494. https://doi.org/10.3233/WOR-2009-0949

Toth, P., Vigo, D., 2002. *The Vehicle Routing Problem*. SIAM, Philadelphia, PA.

Wolsey, L.A., 2020. *Integer Programming*. John Wiley & Sons, Hoboken, NJ.

2 Process Planning

2.1 INTRODUCTION

Process planning covers many disciplines and activities in manufacturing a product or part, from the part's design to the selection of materials and ordering of manufacturing processes. The manufacturing process includes (i) sequencing, (ii) machine selection, (iii) determination of the processing time, and (iv) creation of material flows. The outputs of the process plans are the selected processes, operation sequences, materials and equipment to be used, estimated production time, and the cost of the part (Park & Khoshnevis, 1993; Pirzada, 1991). Process planning is essential in the manufacturing and service sectors. Efficient and carefully prepared process plans will help reduce costs in enterprises and deliver faster and more efficient production (Culler & Burd, 2007; Sutton, 1989). Process plans can be made manually by a planner of small businesses that produce fewer complex products. In this type of traditional process planning, planned jobs are usually assigned to the planner working in the company. Therefore, the performance of a process plan directly depends on the employee's ability to prepare it (Cay & Chassapis, 1997; Marri et al., 1998).

Process planning was presented using a computer-aided structure. Computers have been used to prepare most process plans for more than 40 years. In this section, we provide a brief introduction to the process planning function. In addition, the computer-aided structure of process planning is discussed.

2.2 CLASSICAL PROCESS PLANNING APPROACHES

Before using computers in process planning, manual process plans were used. Part classification and coding processes should be performed correctly to more accurately and efficiently prepare process plans. Therefore, the part classification and coding must be completed before process planning. It is necessary to create part families to classify parts. Part families are created according to the design and manufacturing characteristics of the parts. Part classification refers to the categorization of parts into similar groups. In coding studies, labels (codes) are assigned to the parts. Thus, the properties of the parts were recognized through the codes. In general, the purpose of a coding system is to provide a standard form for similar parts.

Many classification and coding systems have been developed in the literature. These systems start with eye control and proceed to complex numerical systems. Businesses can use coding and classification systems. Some well-known coding systems in the literature are given below:

Metal Institute Classification System (MICLASS): The MICLASS system contains 12-digit codes (Patel, 1991). The 12 digits and their means are shown in Figure 2.1.

DOI: 10.1201/9781003215295-2

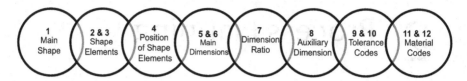

FIGURE 2.1   MICLASS coding system.

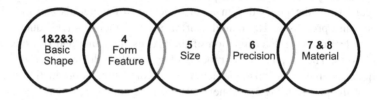

FIGURE 2.2   DCLASS coding system.

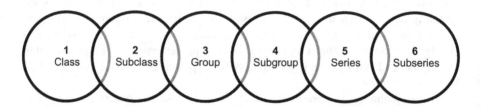

FIGURE 2.3   Brisch coding system.

DCLASS System: Developed by Allen Dell K in 1979. DCLASS is given in a structured manner; it keeps the material's components, processes, and machines in eight characters. The contents of these eight characters are presented in Figure 2.2 (Allen, 1979).

Brisch System: Developed by Brisch. It is one of the oldest coding systems. Figure 2.3 summarizes the coding system as a 9-character-long array (Gombinski, 1968).

Other coding systems can be expressed as OPITZ (Opitz & Wiendahl, 1971), JCODE (Chang et al., 1985), and FORCOD systems (Jung & Ahluwalia, 1991) which are less commonly used in the industry. Part classification and coding systems are a subject within group technology (GT) that benefits the creation of part families and the determination of machine groups, saving time and effort (Halevi & Weill, 1995; Mitrofanov, 1966).

2.3   COMPUTER-AIDED PROCESS PLANNING

Computer-aided process planning (CAPP) means that computers create multiple process plans (Engelke, 1987). With CAPP systems, computer-aided design (CAD) and computer-aided manufacturing (CAM) technologies have become possible.

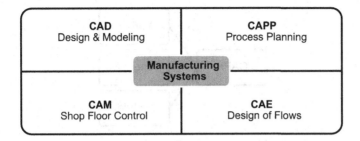

FIGURE 2.4 Relationship between automated manufacturing systems.

Modeling of 3D objects with the help of a computer was carried out using CAD technology (Sarcar et al., 2008; Verroust et al., 1992). In CAD systems, the part's information is entered into a computer database, and the computer converts it to how the machine can process it (Crow, 1992). In cases where computers were not in use, it took much time to create the process plans. In addition, such process plans can be evaluated as low quality. To produce a product or part, to determine, planning, and managing the processes, machinery, and equipment used in the processes will increase production quality. CAPP applications must be developed using computer-integrated manufacturing (CIM) technologies. An improved approach to the design of features in CIM technology is known as automatic feature recognition (Turley et al., 2014). The computer usage areas for automated manufacturing systems are shown in Figure 2.4.

Rehg (1995) proposed three steps for implementing a CAPP system. The steps are as follows:

Step 1: Evaluation of the institution in terms of technology, human resources, and systems

First, the institution's SWOT analysis is conducted, and the system's current status is evaluated.

Step 2: Simplification and elimination of unnecessary transactions

In the second step, it will increase productivity and make production faster by eliminating unnecessary processes. Unnecessary processes are those that do not add value to a product.

Step 3: Implement performance criteria

In the third step, the performance of the installed system is measured. Production systems include performance indicators such as product cycle times, preparation times, process times, manufacturing efficiency, and quality. After the system is installed, continuous improvement in performance indicators is considered the primary goal.

With these steps, part coding and classification studies are the initial stages of a good CAPP system. Classification and coding systems determine the parts with similar characteristics. Parts of the same class are expected to have more common

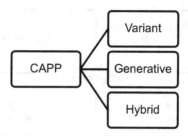

FIGURE 2.5   CAPP systems.

process plans. For a part to be entered into the CAPP system, it is necessary to know its process characteristics, which part its family has, which machines it must be treated on, and what operations it will undergo. Modifications can be made after the process plans are created in CAPP systems, or existing plans can be removed. CAPP technology has enabled automation in production, faster use and coordination of machines and robots, and predictable production. With these opportunities, integrated manufacturing systems have emerged because of CAPP technologies. Chapter 5 provides information about integrated process planning and scheduling (IPPS). For IPPS, the problem in the literature is preferred as the best-performing pair between alternative process plans and different combinations of production scheduling. Thus, the production master plans are facilitated.

With CAPP, it is easier to create process plans automatically. Thus, an opportunity to create more accurate, high-quality, and efficient process plans was achieved. The CAPP systems are approximately handled using three approaches, as shown in Figure 2.5. These are variant, generative, and hybrid approaches that are increasingly being used today (Halevi & Weill, 1995). In the generative process planning method, the process planning phase is initiated by using the definitions of the parts. This approach involves a process without human intervention. Variant process planning is based on a production plan implemented in the past. The production plan is revised, and the new production plan is prepared. During the previous process plans revision, similarities between the parts were reassessed. Consequently, the process is completed with a new plan in which similar parts are considered.

Alternative process plans were created based on the CAPP systems. In traditional production systems, only a single-process plan is prepared. The scheduling function can be operated based on the preparation process plan. Global production performance can be impaired by process planning and scheduling functions that operate independently. Thanks to integrated systems creating more than one single plan (multiple alternative process plans) and the possibility of comparing different scheduling plans, enterprises have increased their products globally.

2.4   REVIEW OF THE LITERATURE

Today, process plans are prepared using computers in small- and large-scale manufacturing enterprises. Therefore, a literature review was initiated using computer-aided process planning studies. The first CAPP studies were conducted in an American business in the early 1970s (Pirzada, 1991). The idea of using computers for process planning was first proposed by Niebel (1965). In recent years, many enterprises have used CAPP applications owing to the widespread use of computers. Parts must be grouped to create computerized process plans (Kumar & Rajotia, 2003).

When CAPP studies are reviewed in the literature, it can be seen that there are too many studies on this topic. The "Computer-Aided process planning" term was searched in the Scopus database. The number of articles published between 1965 and 2021 was 8.423. When we look at the annual distribution of articles, we see that the average number of articles increased in the 2000s. The most published country was the United States, with 2,036 articles, followed by China, which takes the second place with 1,023 articles. The CAPP publication numbers over the years and the publication numbers of the countries are shown in Figures 2.6 and 2.7, respectively.

When looking at the journals in which the articles were published, the most published journal is the *International Journal of Advanced Manufacturing Technology*, which publishes 229 articles. Following this journal, *Proceedings of the ASME Design Engineering Technical Conference* came second, with 183 publications (Table 2.1).

Because the number of articles published on CAPP is too high, there is a need to publish a review article on this subject. The first review article was published in 1984 (Steudel, 1984). In that article, computer-aided systems were examined to establish a connection between design and production. In the following years, the number of articles published on CAPP increased rapidly. In 1989, another review article was published, in which 156 articles were evaluated (Alting & Zhang, 1989). In the 2000s, CAPP increased with the proliferation of computers and the emergence of products with more complex process plans in the industry. In 1997, Cay and Chassapis (1997) reviewed several articles and presented studies on the future of CAPP technologies.

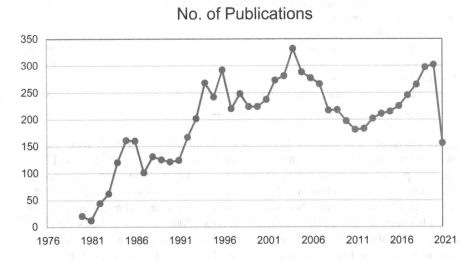

FIGURE 2.6 Number of publications on CAPP between 1970 and 2021.

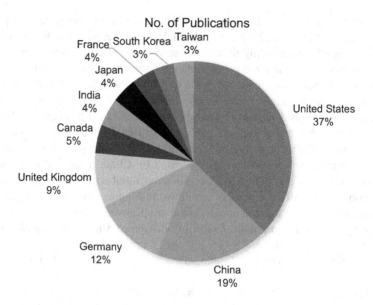

FIGURE 2.7   Publications by countries between 1970 and 2021.

TABLE 2.1

Number of Publications in Journals between 1970 and 2021

Publication	# of Articles
International Journal of Advanced Manufacturing Technology	229
Proceedings of the ASME Design Engineering Technical Conference	183
International Journal of Production Research	164
Computer Integrated Manufacturing Systems CIMS	142
Lecture Notes in Computer Science, Including Subseries Lecture Notes in Artificial Intelligence and Lecture Notes in Bioinformatics	128
Computers and Industrial Engineering	114
CIRP Annals Manufacturing Technology	107
Computers in Industry	102
Journal of Materials Processing Technology	93
Proceedings of SPIE the International Society for Optical Engineering	84

A more recent review article that Yusof and Latif (2014) published, titled "Survey on Computer-Aided Process Planning," reviewed the articles between 2002 and 2013. The first publication examined by the review was an article in 1965 that proposed computers for process planning. Twelve articles were reviewed before the 2000s. A total of 105 articles were then included in the review, including 58 articles published between 2002 and 2007 and 35 articles from 2008 to 2013. Articles are classified based on their methods: feature-based technologies, artificial neural networks, genetic algorithms, fuzzy clusters and logic, Petri nets, factor-based systems, and internet-based systems.

TABLE 2.2
Approaches Used to Optimize CAPP

Algorithms	References
Genetic Algorithms	Li et al. (2005); Ma et al. (2002); Reddy (1999); Salehi and Tavakkoli-Moghaddam (2009); Su et al. (2018); Váncza and Márkus (1991)
Artificial Neural Networks	Chang and Chang (2000); Devireddy (1999); Lian et al. (2012); Rana et al. (2013); Rojek (2010); Wang et al. (2012)
Genetic Algorithm and Simulated Annealing	Li et al. (2002)
Imperialist Competitive Algorithm	Lian et al. (2012)
Artificial Immune System	Prakash et al. (2012)
Honeybees Mating Optimization	Wen et al. (2014)

Another review article published in 2018, titled "A Survey on Smart, Automated Computer-Aided Process Planning (ACAPP) Techniques" (Al-Wswasi et al., 2018), aims to obtain more accurate and automation-based results with automatic computer-aided process planning (ACAPP), an advanced level of CAPP technologies.

Many methods have been developed for optimizing CAPP for selecting manufacturing operations and machines. Genetic algorithms are among the most commonly used methods (Xu et al., 2011). Table 2.2 summarizes the articles based on the methods used by Grabowik et al. (2013) and Isnaini and Shirase (2014).

2.5 CONCLUSION

CAPP systems play a crucial role in CAD/CAM technologies. With CAPP systems, the process of turning parts into final products is determined. One of the most critical outcomes of process planning is the selection and sequencing of operations. Operation sequences show the order of machines to which a part will be operated and are called job routes. CAPP systems can define alternative job routes. Process planning and scheduling functions must be performed together to select the best route for entering the scheduling function among alternative routes. Process planning and scheduling are two essential functions of production shop floors. Because significant gains will be achieved by combining these two functions, integrated studies have been emphasized. These studies were conducted under integrated process planning and scheduling (IPPS). Later sections of the book discuss studies on the integration of production functions.

REFERENCES

Allen, D. K. (1979). *Generative process planning using the DCLASS Information System.* Computer Aided Manufacturing Laboratory Brigham Young University.
Alting, L., & Zhang, H. (1989). Computer aided process planning: The state-of-the-art survey. *The International Journal of Production Research*, 27(4), 553–585.

Al-Wswasi, M., Ivanov, A., & Makatsoris, H. (2018). A survey on smart automated computer-aided process planning (ACAPP) techniques. *The International Journal of Advanced Manufacturing Technology, 97*(1), 809–832.

Cay, F., & Chassapis, C. (1997). An IT view on perspectives of computer aided process planning research. *Computers in Industry, 34*(3), 307–337.

Chang, P.-T., & Chang, C.-H. (2000). An integrated artificial intelligent computer-aided process planning system. *International Journal of Computer Integrated Manufacturing, 13*(6), 483–497. https://doi.org/10.1080/09511920050195922

Chang, Y., Zhu, K., Wu, G., Wong, D. F., & Wong, C. K. (1985). *An introduction to automated process planning*. Prentice-Hall.

Crow, K. (1992). *Computer-aided process planning*. DRM Associates.

Culler, D. E., & Burd, W. (2007). A framework for extending computer aided process planning to include business activities and computer aided design and manufacturing (CAD/CAM) data retrieval. *Robotics and Computer-Integrated Manufacturing, 23*(3), 339–350. https://doi.org/10.1016/j.rcim.2006.02.005

Devireddy, C. R. (1999). Feature-based modelling and neural networks based CAPP for integrated manufacturing. *International Journal of Computer Integrated Manufacturing, 12*(1), 61–74.

Engelke, W. D. (1987). *How to integrate CAD/CAM systems: Management and technology*. CRC Press.

Gombinski, J. (1968). Component classification-why and how? *Summary of a Paper Presented in Machinery and Production Engineering*, 547–550.

Grabowik, C., Kalinowski, K., Kempa, W. M., & Paprocka, I. (2013). A survey on CAPP systems development methods. *Advanced Materials Research, 837*, 387–392. https://doi.org/10.4028/www.scientific.net/AMR.837.387

Halevi, G., & Weill, R. (1995). *Principles of process planning: A logical approach*. Springer, the Netherlands. https://doi.org/10.1007/978-94-011-1250-5

Isnaini, M. M., & Shirase, K. (2014). Review of computer-aided process planning systems for machining operation–future development of a computer-aided process planning system. *International Journal of Automation Technology, 8*(3), 317–332.

Jung, J.-Y., & Ahluwalia, R. S. (1991). FORCOD: A coding and classification system for formed parts. *Journal of Manufacturing Systems, 10*(3), 223–232.

Kumar, M., & Rajotia, S. (2003). Integration of scheduling with computer aided process planning. *Journal of Materials Processing Technology, 138*(1–3), 297–300. https://doi.org/10.1016/S0924-0136(03)00088-8

Li, L., Fuh, J. Y. H., Zhang, Y. F., & Nee, A. Y. C. (2005). Application of genetic algorithm to computer-aided process planning in distributed manufacturing environments. *Robotics and Computer-Integrated Manufacturing, 21*(6), 568–578.

Li, W. D., Ong, S. K., & Nee, A. Y. C. (2002). Hybrid genetic algorithm and simulated annealing approach for the optimization of process plans for prismatic parts. *International Journal of Production Research, 40*(8), 1899–1922.

Lian, K., Zhang, C., Shao, X., & Gao, L. (2012). Optimization of process planning with various flexibilities using an imperialist competitive algorithm. *The International Journal of Advanced Manufacturing Technology, 59*(5–8), 815–828.

Ma, G. H., Zhang, F., Zhang, Y. F., & Nee, A. Y. C. (2002). An automated process planning system based on genetic algorithm and simulated annealing. *International Design Engineering Technical Conferences and Computers and Information in Engineering Conference, 36231*, 57–63.

Marri, H. B., Gunasekaran, A., & Grieve, R. J. (1998). Computer-aided process planning: A state of art. *The International Journal of Advanced Manufacturing Technology, 14*(4), 261–268.

Mitrofanov, S. P. (1966). *The Scientific Principles of Group Technology, Volumes I, II & III*. National Lending Library for Science and Technology.

Niebel, B. W. (1965). Mechanized process selection for planning new designs. *ASME Paper, 737*.

Opitz, H., & Wiendahl, H.-P. (1971). Group technology and manufacturing systems for small and medium quantity production. *The International Journal of Production Research*, *9*(1), 181–203.

Park, J. Y., & Khoshnevis, B. (1993). A real-time computer-aided process planning system as a support tool for economic product design. *Journal of Manufacturing Systems, 12*(2), 181–193. https://doi.org/10.1016/0278-6125(93)90017-N

Patel, S. K. (1991). *Development of a Classification and Coding System for Computer-Aided Process Planning* [Master of Science, New Jersey Institute of Technology]. https://digital commons.njit.edu/theses/1287

Pirzada, S. (1991). *Computer-Aided Process Planning (CAPP) SORICH* [Master of Science, New Jersey Institute of Technology]. https://digitalcommons.njit.edu/theses/1288.

Prakash, A., Chan, F. T. S., & Deshmukh, S. G. (2012). Application of knowledge-based artificial immune system (KBAIS) for computer aided process planning in CIM context. *International Journal of Production Research, 50*(18), 4937–4954.

Rana, A. S., Kumar, R., Singh, M., & Kumar, A. (2013). Operation sequencing in CAPP by using artificial neural network. *International Journal of Innovative Research in Science, Engineering and Technology, 2*(4), 1137–1141.

Reddy, S. B. (1999). Operation sequencing in CAPP using genetic algorithms. *International Journal of Production Research, 37*(5), 1063–1074.

Rehg, J. A. (1995). *Computer-Integrated Manufacturing*. Prentice-Hall, Inc.

Rojek, I. (2010). Neural networks as performance improvement models in intelligent CAPP systems. *Control and Cybernetics, 39*, 54–68.

Salehi, M., & Tavakkoli-Moghaddam, R. (2009). Application of genetic algorithm to computer-aided process planning in preliminary and detailed planning. *Engineering Applications of Artificial Intelligence, 22*(8), 1179–1187. https://doi.org/10.1016/j.engappai.2009.04.005

Sarcar, M. M. M., Rao, K. M., & Narayan, K. L. (2008). *Computer Aided Design and Manufacturing*. PHI Learning Pvt. Ltd.

Steudel, H. J. (1984). Computer-aided process planning: Past, present and future. *The International Journal of Production Research, 22*(2), 253–266.

Su, Y., Chu, X., Chen, D., & Sun, X. (2018). A genetic algorithm for operation sequencing in CAPP using edge selection based encoding strategy. *Journal of Intelligent Manufacturing, 29*(2), 313–332.

Sutton, G. P. (1989). Survey of process planning practices and needs. *Journal of Manufacturing Systems, 8*(1), 69–71.

Turley, S. P., Diederich, D. M., Jayanthi, B. K., Datar, A., Ligetti, C. B., Finke, D. A., Saldana, C., & Joshi, S. (2014). Automated process planning and CNC-code generation. In *IIE Annual Conference. Proceedings*, (p. 2138). Institute of Industrial and Systems Engineers (IISE).

Váncza, J., & Márkus, A. (1991). Genetic algorithms in process planning. *Computers in Industry, 17*(2–3), 181–194. https://doi.org/10.1016/0166-3615(91)90031-4

Verroust, A., Schonek, F., & Roller, D. (1992). Rule-oriented method for parameterized computer-aided design. *Computer-Aided Design, 24*(10), 531–540. https://doi.org/10.1016/0010-4485(92)90040-H

Wang, J., Zhang, H. L., & Su, Z. Y. (2012). Manufacturing knowledge modeling based on artificial neural network for intelligent CAPP. *Applied Mechanics and Materials, 127*, 310–315.

Wen, X., Li, X., Gao, L., & Sang, H. (2014). Honey bees mating optimization algorithm for process planning problem. *Journal of Intelligent Manufacturing, 25*(3), 459–472.

Xu, X., Wang, L., & Newman, S. T. (2011). Computer-aided process planning – A critical review of recent developments and future trends. *International Journal of Computer Integrated Manufacturing*, 24(1), 1–31. https://doi.org/10.1080/0951192X.2010.518632

Yusof, Y., & Latif, K. (2014). Survey on computer-aided process planning. *The International Journal of Advanced Manufacturing Technology*, 75(1–4), 77–89. https://doi.org/10.1007/s00170-014-6073-3

3 Scheduling in Manufacturing

3.1 INTRODUCTION

Many events in a manufacturing or service system require deciding on an option within many alternatives, which is called decision-making. A crucial decision-making process is scheduling for these systems, which interact with other decision-making functions. Scheduling is an inseparable part of manufacturing and is studied intensively in academia as it significantly impacts the manufacturing sector. Even a slight gain in this system may save considerable money and time.

Scheduling is a fundamental decision-making process used in most manufacturing and service industries. It can be defined as assigning resources to tasks in given time periods, aiming to optimize one or more objectives (Pinedo, 2012). In modern factories, scheduling is performed within an enterprise-wide information system. Data is collected from personal computers to data entry terminals and processed using enterprise resource planning systems. With the fourth industrial revolution (Industry 4.0), collecting data and processing it in real time to use in scheduling is indeed providing new possibilities for development and attracting researchers on this topic (Fernandez-Viagas and Framinan, 2022).

Scheduling can be affected by unexpected events on the shop floor, such as machine breakdowns, delayed processes, and occupational work accidents. To control the operations and maintain efficiency, it is required to prepare detailed task schedules and manage them accordingly. The scheduling relationship with other parts of the organization is shown in Figure 3.1 (Pinedo, 2012).

3.2 REVIEW OF THE LITERATURE

Scheduling in manufacturing has been intensively studied. The first studies on this topic can be found in the 1950s (Salveson, 1952; Johnson, 1954; Bowman, 1959). We will provide the latest studies in this area to indicate the current status. Bibliometric information from the 7,981 articles indexed in the science citation index–expanded (SCI-Exp) are obtained from the Web of Science Core Collection library using the keywords "scheduling" and "shop" between the years 1981 and 2022, written in the English language. Most (5,970) of these studies are from the Web of Science categories of the "Operations Research Management Science," "Engineering Industrial," "Engineering Manufacturing," "Computer Science Interdisciplinary Applications," and "Computer Science Artificial Intelligence" areas. This data is visualized using the Biblioshiny interface of Bibliometrix software, which is based on the R programming language (Aria and Cuccurullo, 2017). Although in the years 2001, 2007, 2010, and 2014 publication numbers decreased compared to the previous years, it is seen

DOI: 10.1201/9781003215295-3

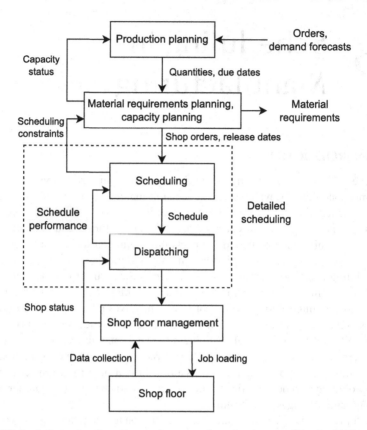

FIGURE 3.1   Information flow diagram in a manufacturing system. (Pinedo, 2012.)

from Figure 3.2 that there is an annual growth rate of 9.85%. There is an increasing trend in the number of studies in this field.

Table 3.1 provides the journals with the most published articles and their impacts on the literature. Two of the most influential metrics for scientific research are h and g indexes, which we will provide in this table (Harzing, 2007). Although most articles are published in the *International Journal of Production Research*, the *European Journal of Operational Research* is the most impactful in the h and g indexes.

In Table 3.2, the most productive authors are listed. Wang L takes first place with 108 publications, 1,931 local citations (17.88 average local citations per article), 5,363 total citations (49.66 average total citations per article), 45 h-index value, and 71 g-index value in our search terms. Local citations show the citations received from articles in the analyzed collection. Global citations, on the other hand, indicate the number of citations received from the entire Web of Science database.

When most cited papers are investigated, the study of Hall and Sriskandarajah (1996) has 278 local and 608 global citations and a 47.72% local-to-global citation ratio with their paper titled "A Survey of Machine Scheduling Problems with Blocking and No-Wait in Process," as seen in Table 3.3. It can be concluded that review articles and studies that suggest new methods have more attention.

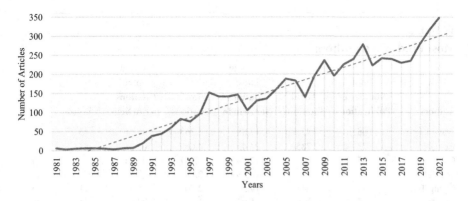

FIGURE 3.2 Number of articles about "scheduling" and "shop" between 1981 and 2021.

TABLE 3.1
Journals with the Most Published Articles and Their Impacts

Journals	No. of Articles	h-index	g-index
International Journal of Production Research	790	58	80
Computers & Industrial Engineering	501	59	93
International Journal of Advanced Manufacturing Technology	402	49	68
European Journal of Operational Research	390	72	131
Computers & Operations Research	345	60	97
International Journal of Production Economics	211	47	72
Expert Systems with Applications	164	39	61
Journal of Intelligent Manufacturing	163	42	63
Applied Soft Computing	132	36	54
Journal of Scheduling	128	27	56
Production Planning & Control	127	26	36
Journal of the Operational Research Society	123	30	45
Annals of Operations Research	119	29	46
International Journal of Computer Integrated Manufacturing	99	23	39
Journal of Manufacturing Systems	97	30	48

The co-occurrence of the 20 most frequently used KeyWords Plus (Garfield and Sher, 1993) words and author keywords, along with years, is visualized using the VOSviewer software and given in Figures 3.3 and 3.4, respectively (van Eck and Waltman, 2010). It is seen that while "optimization" and "particle swarm optimization" have been used relatively more recently in KeyWord Plus words, "multi-objective" and "energy" keywords are used by authors in recent studies.

The 50 most used trigrams in article titles are visualized in Figure 3.5 to depict the current studies. It is seen that "job shop scheduling," "flow shop scheduling," and "flexible job shop" is immensely used in large part of the studies.

TABLE 3.2
Most Productive Authors

Authors	Total Number of Publications	Local Citations	Average Local Citations per Article	Total Citations	Average Total Citations per Article	h-index	g-index
Wang L	108	1931	17.88	5,363	49.66	45	71
Pan QK	73	1785	24.45	4,669	63.96	44	68
Gao L	55	953	17.33	2,437	44.31	27	49
Zandieh M	55	1,041	18.93	2,398	43.6	31	48
Cheng TCE	45	647	14.38	2,502	55.6	21	45
Lei DM	45	676	15.02	1,559	34.64	24	39
Wang JB	47	435	9.26	1,596	33.96	27	39
Framinan JM	40	925	23.13	1,826	45.65	23	40
Li JQ	39	699	17.92	1,980	50.77	27	39
Werner F	42	489	11.64	1,458	34.71	22	38

TABLE 3.3
Most Cited Papers

Author(s) and Publication Year	Title	LC	GC	LC/GC	NLC	NGC
Hall and Sriskandarajah (1996)	A Survey of Machine Scheduling Problems with Blocking and No-Wait in Process	278	608	45.72	17.47	5.45
van Laarhoven et al. (1992)	Job Shop Scheduling by Simulated Annealing	249	617	40.36	16.21	13.47
Nowicki and Smutnicki (1996)	A Fast Taboo Search Algorithm for the Job Shop Problem	243	589	41.26	15.27	5.28
Ruiz and Vázquez-Rodríguez (2010)	The hybrid flow shop scheduling problem	236	469	50.32	16.53	12.27
Ruiz and Stützle (2007)	A simple and effective iterated greedy algorithm for the permutation flowshop scheduling problem	211	652	32.36	16.48	14.85
Allahverdi et al. (2008)	A survey of scheduling problems with setup times or costs	192	817	23.5	12.33	17.58
Pezzella et al. (2008)	A genetic algorithm for the Flexible Job-shop Scheduling Problem	191	529	36.11	12.27	11.38
Kacem et al. (2002)	Approach by localization and multi-objective evolutionary optimization for flexible job-shop scheduling problems	180	425	42.35	16.9	12.08
Jain and Meeran (1999)	Deterministic job-shop scheduling: Past, present and future	163	440	37.05	11.49	8.6
Xia and Wu (2005)	An effective hybrid optimization approach for multi-objective flexible job-shop scheduling problems	162	431	37.59	14.05	11.98

LC: Local Citations, GC: Global Citations, LC/GC: LC/GC Ratio (%), NLC: Normalized Local Citations, NGC: Normalized Global Citations.

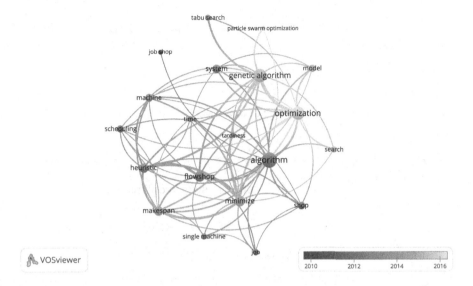

FIGURE 3.3 Co-occurrence of KeyWords Plus keywords along with the years.

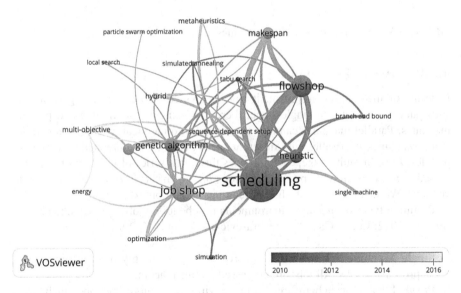

FIGURE 3.4 Co-occurrence of author keywords along with the years.

3.3 SCHEDULING ENVIRONMENTS

Although there are several notations available for representing scheduling environments, we can define scheduling environments with $\alpha \mid \beta \mid \gamma$ notation. The term α represents the machine environment; β shows the processing characteristics; γ indicates the minimization objective (Pinedo, 2012). A visual representation of scheduling environments can be seen in Figure 3.6 (Đurašević and Jakobović, 2022).

FIGURE 3.5   Most used trigrams in article titles.

3.3.1 Machine Environment

In terms of machine environment, it can be divided into two sub-stages: single-stage and multi-stage. In single-stage, there exist single machine (1) and parallel machines. Parallel machines can also be divided into identical machines in parallel (Pm), machines in parallel with different speeds (Qm), and unrelated machines in parallel (Rm). In multi-stage, each job goes through a series of machines to be processed. There exist flow shop (Fm), job shop (Jm), open shop (Om), and their flexible versions. We will focus on job shop scheduling in our context.

Common types of machine environments in scheduling jobs are explained below (Pinedo, 2012; Li and Gao, 2020; Đurasević and Jakobović, 2022):

Single Machine Scheduling: This problem type has only one machine for processing jobs. All jobs are processed on this machine.

Parallel Machine Scheduling: In this environment, more than one machine is available for processing jobs. There are three kinds of parallel machine scheduling based on the processing speed of machines. In the first one, there are several machines with the same (identical) processing properties (Pm). A job can be processed in any one of them. Machines produce the same output but with different processing speeds (Qm) in the second one. Jobs have different processing speeds in different parallel machines (Rm) in the third one, which is a more general form of this machine environment.

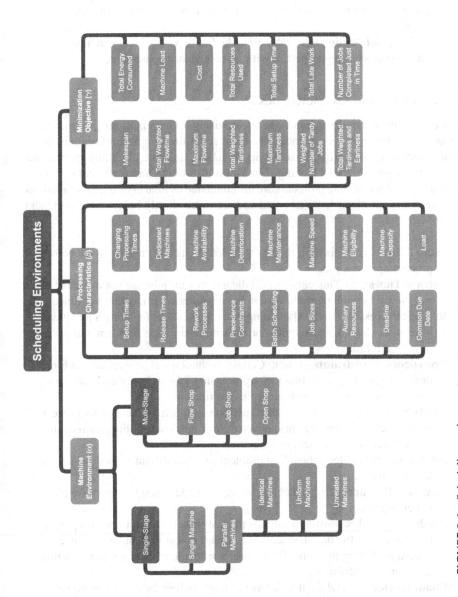

FIGURE 3.6 Scheduling environments.

Job Shop Scheduling (*Jm*): In this problem, there are "*m*" machines with different processing properties, and each job "*j*" has to be processed along with a specific route to the finish, depending on its precedence constraints.

Flow Shop Scheduling (*Fm*): In this case, there are machines with different processing properties, but each job must be processed with the same route.

Open Shop Scheduling (*Om*): A job can be freely processed in machines with different processing properties without precedence constraints.

3.3.2 PROCESSING CHARACTERISTICS

Processing characteristics can include release dates (r_j), preemptions (*prmp*), precedence constraints (*prec*), sequence-dependent setup times (s_{jk}), job families (*fmls*), batch processing (*batch(b)*), breakdowns (*brkdwn*), machine eligibility restrictions (M_j), permutation (*prmu*), blocking (*block*), no-wait (*nwt*), and recirculation (*rcrc*). As the types of scheduling problems are differentiated as time goes on, new notations to cover those problems have emerged and are given shortly below (Đurasević and Jakobović, 2022).

Setup Times (*s*): A specific time is required to prepare the machines before processing a new job.

Release Times (r_j): This constraint indicates that all jobs are not available at once at the beginning of the system but are released gradually over time.

Rework Processes (*rwrk*): After completing a process or finishing a job, it can have some defects and requires processing again. This problem can occur once or multiple times.

Precedence Constraints (*prec*): Certain technological constraints enforce processing in a specific order. So, processing jobs independently in a random manner is not applicable most of the time.

Batch Scheduling (*batch*): Similar jobs are grouped in batches to process them together. There are various versions of batch scheduling, such as different batch sizes or machine capacities.

Job Sizes (J_s): In batch scheduling problems, this constraint shows the size of jobs in a batch.

Auxiliary Resources (*R*): Indicates the required additional resources for a job to begin processing.

Deadline ($\bar{d}_j$): This constraint implies the last time for a job to be completed. It differs from the due date as this is a hard constraint, and a job must be completed before this time. Otherwise, a schedule without agreeing on this constraint is unfeasible.

Common Due Date (d_j): All jobs have to finish before the same common due date or common deadline.

Changing Processing Times (p_c): This indicates the variations in the processing time of the jobs caused by many reasons, such as using extra resources or an increase in workers learning curve.

Dedicated Machines (M_{ded}): Some jobs may be processed faster in specific machines. This speedup could be possible due to some special property of that machine that other machines do not have.

Machine Availability (*brkdwn*): Due to expected events such as planned maintenance or unexpected events such as machine breakdowns, machines may not always be available.

Machine Deterioration (M_d): After executing a machine for a while, it may become slower due to the wearing of tools, etc., and needs maintenance.

Machine Maintenance (M_m): It needs some maintenance activities to keep the system in working condition. There are two types of maintenance, which are preventive and corrective maintenance.

Machine Speed (M_s): Speed may be adjusted by changing resources for machines, such as energy.

Machine Eligibility (M_j): A job can only be processed on the appropriate machine or machines.

Machine Capacity (M_c): In batch scheduling, machines may have different capacities. Some machines can process larger batches, while others can process smaller batches.

Load (*L*): Jobs that will be processed should be transferred and loaded to the related machines. Transportation can be done with various vehicles with different capacities.

3.3.3 MINIMIZATION OBJECTIVE

Regular performance measures in objective functions to be minimized can be makespan (C_{max}), maximum lateness (L_{max}), total weighted completion time ($\Sigma w_j C_j$), discounted total weighted completion time ($\Sigma w_j (1 - e^{-rC_j})$), total weighted tardiness ($\Sigma w_j T_j$), and the weighted number of tardy jobs ($\Sigma w_j U_j$). There may be more than one objective, which is called multi-objective optimization. Below most used objectives are given (Đurasević and Jakobović, 2022).

Makespan (C_{max}): Latest completion time of the last job in a schedule.

Total Weighted Flowtime (*Fwt*): Time spent by jobs in the system is multiplied by a specific weight and summed together.

Maximum Flowtime (F_{max}): The flowtime of the one with the longest flowtime of all jobs in the system.

Total Weighted Tardiness (*Twt*): The tardiness of each job in the system is multiplied by a specific weight and summed together.

Maximum Tardiness (T_{max}): The tardiness of the one with the longest tardiness of all jobs in the system.

Weighted Number of Tardy Jobs (*Uwt*): Tardy jobs in the system are multiplied by a specific weight and summed together.

Total Weighted Tardiness and Earliness (*Etwt*): Sum of the weighted earliness and tardiness together.

Total Energy Consumed (*TEC*): Total energy consumed between starting and ending the system.

Machine Load (*ML*): Focuses on balancing the loads in machines.

Cost (*COST*): Focuses on costs of system execution.

Total Resources Used (R_u): Focuses on resource usage only if additional resources are consumed in the system.

Total Setup Time (*Ts*): Focuses on time spent on setups.

Total Late Work (*TLW*): Focuses on the amount of remaining processing of all jobs. A job is considered late if it passes its due date.

Number of Jobs Completed Just in Time (N_{jit}): Focuses on the number of jobs that finishes just in time, in which jobs are completed at the exact due dates.

3.3.4 PROBLEM PROPERTIES

When we investigate real-life shop scheduling problems, they have specific characteristics shortly given below (Li and Gao, 2020):

Multi-objective: When scheduling jobs, several different objectives may conflict with each other, such as increasing productivity to deliver products quickly causes a rise in work-in-progress levels, which in turn increases costs (Berry, 1990). So there is more than one objective that should be considered simultaneously in scheduling.

Multi-constraint: Many constraints make it difficult to make a schedule. Starting with jobs, there are process routes, the processing capability of machines, resource constraints, and more.

Uncertainty: The manufacturing environment is open to changes as they seem to occur randomly, such as machine breakdowns, new, changed, or canceled tasks. So it is a dynamic environment that has many uncertainties in nature.

Discretization: The manufacturing system can be modeled with discrete events, such as start times of jobs, new jobs arrival, or machine breakdown. It makes it possible to investigate scheduling problems using mathematical programming, discrete event simulation, and other methods.

Computational complexity: Scheduling is a combinatorial optimization problem in the NP-complete problem complexity class in most cases.

3.4 DISPATCHING RULES

Scheduling algorithms can be classified as exact, approximation, and heuristic algorithms. Heuristic algorithms can be divided into construction and improvement heuristics. Construction heuristics begins without a schedule and append a single job in each iteration. On the other hand, improvement heuristics start with a schedule and search for a better one similar to the one at hand. Although basic dispatching rules are explained in Chapter 10, and solution techniques for the integrated manufacturing functions are given in Chapter 11, here we will provide the general-purpose procedures for deterministic scheduling.

All of the dispatching rules belong to the construction heuristics. Improvement heuristics are local search algorithms such as iterative improvement, threshold accepting, simulated annealing, and tabu search. A dispatching rule is a policy that prioritizes all jobs waiting to be processed on a machine. A dispatching rule explores the queued jobs and picks the job with the highest priority once a machine is available (Pinedo, 2012):

There are many dispatching rules used in scheduling. They can be classified based on dependence on the time in the shop, dependence on the state of the shop, the rules' structure, and the rules' information content (Ramasesh, 1990). Rules that are not time-dependent are also called static, while time-dependent rules are called dynamic. Basic rules can be integrated to obtain more complex rules called composite rules to handle more complicated objective functions. For example, Apparent Tardiness Cost (ATC) combines the WSPT and MS rules. Apparent Tardiness Cost with Setups (ATCS) also integrates SST over ATC. Conditionally Expediting SPT (CEXPST) combines SPT with the expediting strategy.

Some of the dispatching rules used in scheduling are given in Table 3.4. In this table, Time dependency (TD) and State of Shop Dependence (SSD) are provided as High (H) and Low (L). Structure (S) is provided as Simple (S) and Complex (C). Information content of the related rule is given in the last column of this table (Haupt, 1989; Grabot and Geneste, 1994; Pinedo, 2012; Shahzad and Mebarki, 2016).

TABLE 3.4

Some of the Dispatching Rules Used in Scheduling

Rule Name	TD	SSD	S	Information Content
Apparent Tardiness Cost (ATC)	H	L	S	Due date, processing time, and tardiness cost
Apparent Tardiness Cost with Setups (ATCS)	H	H	C	Due date, processing time, tardiness cost, and setup cost
Conditionally Expediting SPT (CEXPST)	H	H	C	Processing time, expediting cost, and priority
Cost Over Time (COVERT)	H	H	C	Processing time, tardiness cost, and priority
Critical Ratio (CR)	H	H	S	Slack time and remaining processing time
Critical Ratio/Shortest Imminent (CRSI)	H	H	C	Slack time, remaining processing time, and priority
Earliest Completion Time (ECT)	H	H	S	Processing time
Earliest Due Date (EDD)	H	L	S	Due date
Earliest Operation Due-Date (ODD)	H	H	S	Due date of the next operation
Earliest Release Date (ERD)	H	H	S	Release date
First Come First Served (FCFS)	H	L	S	Arrival time
Fewest Number of Operations Remaining (FOPNR)	L	H	S	Number of operations remaining
First Arrival at Shop First Served (FASFS)	H	L	S	Arrival time at the shop

(Continued)

TABLE 3.4 (Continued)
Some of the Dispatching Rules Used in Scheduling

Rule Name	TD	SSD	S	Information Content
Greatest Number of Operations Remaining (GOPNR)	L	H	S	Number of operations remaining
Greatest Total Work (TWORK)	L	H	S	Total work remaining
Least Work Remaining (LWKR)	L	H	S	Work remaining
Longest Alternate Processing Time (LAPT)	L	H	S	Alternate processing time
Longest Processing Time (LPT)	L	H	S	Processing time
Longest Total Remaining Processing on Other Machine (LTRPOM)	L	H	S	Remaining processing time on other machines
Minimum Slack (MS)	H	H	S	Slack time
Modified Operation Due Date (MOD)	H	H	S	Due date of the next operation
Most Work Remaining (MWKR)	L	H	S	Work remaining
Multi-Factor (MF)	H	H	C	Multiple factors such as due date, processing time, and priority
Service In Random Order (SIRO)	L	H	S	Random priority
Shortest Imminent Processing (SI)	H	H	S	Imminent processing time
Shortest Processing Time (SPT)	L	H	S	Processing time
Shortest Queue (SQ)	L	H	S	Queue length
Shortest Queue at the Next Operation (SQNO)	L	H	S	Queue length at the next operation
Shortest Setup Time (SST)	L	H	S	Setup time
Slack (SLACK)	H	H	S	Slack time
Smallest Allowance (ALL)	H	L	S	Allowance
Smallest Critical Ratio (CR)	H	H	S	Slack time and remaining processing time
Smallest Operation Critical Ratio (OCR)	H	H	S	Slack time, remaining processing time, and number of operations remaining
Smallest Operation Slack (OSL)	H	H	S	Slack time of the next operation
Smallest Ratio of Allowance per Number of Operations Remaining (ALL/OPN)	H	H	C	Allowance, number of operations remaining, and priority
Smallest Ratio of Slack per Allowance (S/ALL)	H	H	C	Slack time, allowance, and priority
Smallest Ratio of Slack per Number of Operations Remaining (S/OPN)	H	H	C	Slack time, number of operations remaining, and priority
Smallest Ratio of Slack per Work Remaining (S/WKR)	H	H	C	Slack time, work remaining, and priority
The smallest value of a random priority (RANDOM)	L	H	S	Random priority
Weighted Shortest Processing Time (WSPT)	L	H	C	Processing time and priority
With Biggest Weight (WI)	H	H	C	Priority and weight

3.5 JOB SHOP SCHEDULING

Job shop scheduling deals with the scheduling problem of the n jobs that will be processed in the m machines. A job has k processes, which can be processed in one or more machines, and technological constraints must be satisfied in this process. Each job has a specific route to follow. Along this route, a job may be processed in the same machine one time or more than once, which is called recirculation. There are different objectives in job shop scheduling, such as makespan, maximum lateness, total weighted completion time, and number of late jobs. A job shop can be represented with discrete models formulated as a linear or nonlinear program (Pinedo, 2005).

Job shop scheduling problems (JSSP) can be classified into 14 groups, as given in Figure 3.7, which are: deterministic, flexible, static, dynamic, periodic, cyclic, preemptive, no-wait, just-in-time, large-scale, re-entrant, assembly, stochastic, and fuzzy (Abdolrazzagh-Nezhad and Abdullah, 2017; Srsen and Mernik, 2021).

In static JSSP, knowledge of all jobs and machines is available for processing at the beginning of the problem (Qiu and Lau, 2014). On the other hand, customer demand can change, jobs come randomly, processing times can also vary with stochastic nature, unscheduled machine breakdowns, workers' absenteeism, etc. are also considered in dynamic JSSP (Mohan et al., 2019). Operation precedence relationships and processing times are known beforehand in both types of problems.

While processing times of all job operations on each machine are known in the beginning in deterministic JSSP, on the other hand, the real-world industry is free from deterministic variables and constraints. The probability of changes in job arrivals and job processing times can be considered in stochastic JSSP (Tavakkoli-Moghaddam et al., 2005). One or more machines can process the same job in flexible JSSP. Real-world applications have many uncertainties, so fuzzy JSSP is introduced to deal with these problems and prepare more realistic schedules owing to attributes such as fuzzy processing times or fuzzy due dates (Liu et al., 2015).

Iteratively processing jobs in batches is called periodic JSSP. There is no completion time in periodic JSSP. Instead, the number of processed pieces leaving the job

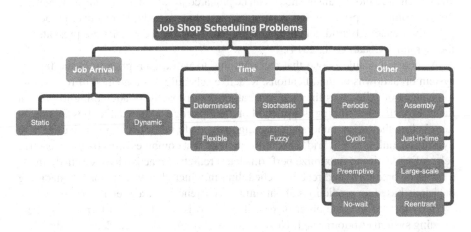

FIGURE 3.7 JSSP classification.

shop in a certain time can be used as a performance metric (Serafini and Ukovich, 1989). Processing all jobs indefinitely often is called cyclic JSSP (Brucker et al., 2012). Each operation has a different penalty for due-date, earliness, and tardiness in just-in-time JSSP (Ahmadian and Salehipour, 2021). Processes of jobs can be interrupted at any time to allow some other processes to execute in preemptive JSSP (Yun, 2002). Real-life manufacturing systems can have many machines (i.e., greater than 100 machines) that can be regarded as large-scale JSSP (Hodgson et al., 1998). Two consecutive job processes cannot be interrupted in no-wait JSSP (Mascis and Pacciarelli, 2002). All jobs go through the same route, and this route is followed a few times to finish jobs in re-entrant JSSP (Chen et al., 2008).

Here in this book, we focus on the dynamic environment in a job shop, so we will give the dynamic job shop as a separate section.

3.6 DYNAMIC JOB SHOP SCHEDULING

Classic JSSP is considered predictive and deliberative, in which a schedule is generated without considering any dynamic event and tried to be executed on the shop floor directly. On the other hand, the execution controller requires some relaxation but cannot make any changes to this fixed schedule (Bidot et al., 2009). In an actual manufacturing environment, many events can affect production severely, such as machine breakdowns, order changes, and operator problems.

Dynamic events can be classified into four groups based on their related factor (Mohan et al., 2019). There can be workpiece-related events such as changing processing times, random workpiece entrances, changing delivery dates, and changing the priority of the orders. There are also machine-related events, such as machine breakdowns and changing machine utilization. Another one is process-related events such as postponed processes, unqualified processes, and some processes that corrupt the production flow. Lastly, other events include absenteeism, feedstock delays, faulty materials, etc.

In dynamic job shop scheduling, random events are included to represent these changes; thus, a more realistic model can be prepared in such an uncertain environment. To overcome this problem and generate more robust schedules, reactive, proactive, proactive-reactive hybrid, and predictive-reactive hybrid approaches are presented in the literature (Chaari et al., 2014).

A robust schedule is one that can continue to work appropriately in the face of system breakdowns or modifications. Reactive scheduling is a method of responding to changes or failures as they happen, rather than proactively attempting to prevent them. Proactive scheduling is a method of anticipating and preventing possible problems before they arise, rather than reacting to them after they occur. Proactive-reactive hybrid scheduling is a hybrid scheduling strategy that combines proactive and reactive scheduling tactics to maximize performance. Predictive-reactive hybrid scheduling is a blend of predictive and reactive scheduling in which the system makes scheduling decisions based on predictions about future occurrences (Cardin et al., 2017).

In dynamic JSSP, the order or priority of jobs is decided, so it turns out to be a queuing system (Abdolrazzagh-Nezhad and Abdullah, 2017). The details of dynamic job shop scheduling are presented in Chapter 6.

3.7 WEIGHTING THE CUSTOMERS

Ideally, meeting customer needs by giving close due dates and keeping promises for all customers is desired. But this ideal could not be realized due to the capabilities of a firm. Even if there is enough capacity to carry out, many uncertainties occur naturally at all supply, production, and delivery phases. Within such an environment giving some priorities to customers based on their value to the company is a solution for getting the most out of them and satisfying the customer needs as optimally as possible. In this case, relatively more important customers can be given closer due dates, while relatively less important customers can get longer due dates. In this way, a company can make the most profitable decision based on the relative importance of its customers and keep its promises more effectively. Demir et al. (2016), Erden et al. (2018), and Demir and Erden (2020) applied this approach and weighted the customers when they assigned due dates in their studies. Also, Güçdemir and Selim (2018) emphasize the importance of customer centricity and reveal that customer-oriented dispatching rules provide better results. Details on weighing the customers are given in Chapter 4.

3.8 COMPUTATIONAL COMPLEXITY OF SCHEDULING

Computational complexity can be categorized into two based on solution time: P (polynomial time) if an optimum solution can be obtained in a polynomial time and NP (non-deterministic Polynomial time) if an optimum solution cannot be obtained in a polynomial time. Most scheduling problems fall into NP-Hard or NP-Complete class problems due to their complexity, so their optimum solutions cannot be found using exact solution techniques. On the other hand, exact solutions can still help obtain optimal solutions for smaller versions of these problems. Also, they can be used to compare the results of heuristic methods in a benchmark.

This example can be given to understand small-large terms better: m-machine flow shop scheduling problem for $m \geq 3$ is in the NP-Complete problem complexity class. As for m-machine, JSSP for $m \geq 2$ is also NP-Complete (Garey et al., 1976). Exact methods can solve small-sized problems within a reasonable time frame.

3.9 CONCLUSION

Scheduling is a decision-making process and an inseparable part of manufacturing, which is briefly discussed in this chapter. There are a considerable number of studies on this topic, and more are coming as new constraints, manufacturing technologies, and solution methods emerge. Companies should seek new appropriate solution methods that will increase their effectiveness and efficiency in their production and compete with other competitors in this harsh competitive environment.

With the recent technological advances known as Industry 4.0, it is now possible to obtain shop floor data and downstream and upstream processes that directly affect the scheduling. So these data should also be considered when preparing schedules. As a result, integrating manufacturing functions such as scheduling with the due-date assignment or process planning is necessary for the companies. Other integrations involve coordinating the production and distribution of goods in a supply chain, such as the integration of production and outbound distribution scheduling. Moreover, determining the optimal production schedule for each product and the most efficient transportation route for the finished products to reach their destination is called the integration of production and transportation scheduling. As for researchers, it is essential to follow up on these novelties to develop new studies in this area.

REFERENCES

Abdolrazzagh-Nezhad, M., Abdullah, S., 2017. Job shop scheduling: Classification, *Constraints and Objective Functions* 11, 6.

Ahmadian, M.M., Salehipour, A., 2021. The just-in-time job-shop scheduling problem with distinct due-dates for operations. *Journal of Heuristics* 27, 175–204. https://doi.org/10.1007/s10732-020-09458-6

Allahverdi, A., Ng, C.T., Cheng, T.E., Kovalyov, M.Y., 2008. A survey of scheduling problems with setup times or costs. *European Journal of Operational Research* 187, 985–1032.

Aria, M., Cuccurullo, C., 2017. bibliometrix: An R-tool for comprehensive science mapping analysis. *Journal of Informetrics* 11, 959–975. https://doi.org/10.1016/j.joi.2017.08.007

Berry, P.M., 1990. Satisfying conflicting objectives in factory scheduling, in: *Sixth Conference on Artificial Intelligence for Applications*, pp. 101–107, vol. 1. https://doi.org/10.1109/CAIA.1990.89177

Bidot, J., Vidal, T., Laborie, P., Beck, J.C., 2009. A theoretic and practical framework for scheduling in a stochastic environment. *Journal of Scheduling* 12, 315–344. https://doi.org/10.1007/s10951-008-0080-x

Bowman, E.H., 1959. The schedule-sequencing problem. *Operations Research* 7, 621–624.

Brucker, P., Burke, E.K., Groenemeyer, S., 2012. A mixed integer programming model for the cyclic job-shop problem with transportation. *Discrete Applied Mathematics* 160, 1924–1935. https://doi.org/10.1016/j.dam.2012.04.001

Cardin, O., Trentesaux, D., Thomas, A., Castagna, P., Berger, T., Bril El-Haouzi, H., 2017. Coupling predictive scheduling and reactive control in manufacturing hybrid control architectures: State of the art and future challenges. *Journal of Intelligent Manufacturing* 28, 1503–1517. https://doi.org/10.1007/s10845-015-1139-0

Chaari, T., Chaabane, S., Aissani, N., Trentesaux, D., 2014. Scheduling under uncertainty: Survey and research directions, in: *2014 International Conference on Advanced Logistics and Transport (ICALT)*, pp. 229–234. https://doi.org/10.1109/ICAdLT.2014.6866316

Chen, J.-S., Pan, J.C.-H., Lin, C.-M., 2008. A hybrid genetic algorithm for the re-entrant flow-shop scheduling problem. *Expert Systems with Applications* 34, 570–577. https://doi.org/10.1016/j.eswa.2006.09.021

Demir, H.I., CaNar, T., Cil, I., Dugenci, M., Erden, C., 2016. Integrating process planning, WMS dispatching, and WPPW weighted due date assignment using a genetic algorithm. *International Journal of Computer and Information Engineering* 10, 1324–1332.

Demir, H.I., Erden, C., 2020. Dynamic integrated process planning, scheduling and due-date assignment using ant colony optimization. *Computers & Industrial Engineering* 149, 106799.

Đurasević, M., Jakobović, D., 2022. Heuristic and metaheuristic methods for the parallel unrelated machines scheduling problem: A survey. *Artificial Intelligence Review*. https://doi.org/10.1007/s10462-022-10247-9

Erden, C., Demir, H.İ., Göksu, A., Uygun, Ö., 2018. Solving process planning, ATC scheduling and due-date assignment problems concurrently using genetic algorithm for weighted customers. *Akademik Platform Mühendislik ve Fen Bilimleri Dergisi* 6, 87–96.

Fernandez-Viagas, V., Framinan, J.M., 2022. Exploring the benefits of scheduling with advanced and real-time information integration in Industry 4.0: A computational study. *Journal of Industrial Information Integration* 27, 100281. https://doi.org/10.1016/j.jiii.2021.100281

Garey, M.R., Johnson, D.S., Sethi, R., 1976. The complexity of flowshop and jobshop scheduling. *Mathematics of Operations Research* 1, 117–129.

Garfield, E., Sher, I.H., 1993. Key words plus [TM]-algorithmic derivative indexing. *Journal-American Society For Information Science* 44, 298–298.

Grabot, B., Geneste, L., 1994. Dispatching rules in scheduling dispatching rules in scheduling: A fuzzy approach. *International Journal of Production Research* 32, 903–915. https://doi.org/10.1080/00207549408956978

Güçdemir, H., Selim, H., 2018. Integrating simulation modelling and multi criteria decision making for customer focused scheduling in job shops. *Simulation Modelling Practice and Theory* 88, 17–31. https://doi.org/10.1016/j.simpat.2018.08.001

Hall, N.G., Sriskandarajah, C., 1996. A survey of machine scheduling problems with blocking and no-wait in process. *Operations Research* 44, 510–525. https://doi.org/10.1287/opre.44.3.510

Harzing, A.W., 2007. Metrics: h and g-index [WWW Document]. Harzing.com. URL https://harzing.com/resources/publish-or-perish/tutorial/metrics/h-and-g-index (accessed 9.23.22).

Haupt, R., 1989. A survey of priority rule-based scheduling. *OR Spektrum* 11, 3–16. https://doi.org/10.1007/BF01721162

Hodgson, T.J., Cormier, D., Weintraub, A.J., Zozom, A., 1998. Note. Satisfying due dates in large job shops. *Management Science* 44, 1442–1446. https://doi.org/10.1287/mnsc.44.10.1442

Jain, A.S., Meeran, S., 1999. Deterministic job-shop scheduling: Past, present and future. *European Journal of Operational Research* 113, 390–434. https://doi.org/10.1016/S0377-2217(98)00113-1

Johnson, S.M., 1954. Optimal two- and three-stage production schedules with setup times included. *Naval Research Logistics Quarterly* 1, 61–68. https://doi.org/10.1002/nav.3800010110

Kacem, I., Hammadi, S., Borne, P., 2002. Approach by localization and multiobjective evolutionary optimization for flexible job-shop scheduling problems. *IEEE Transactions on Systems, Man, and Cybernetics, Part C (Applications and Reviews)* 32, 1–13. https://doi.org/10.1109/TSMCC.2002.1009117

Li, X., Gao, L., 2020. *Effective Methods for Integrated Process Planning and Scheduling, Engineering Applications of Computational Methods.* Springer-Verlag, Berlin and Heidelberg. https://doi.org/10.1007/978-3-662-55305-3

Liu, B., Fan, Y., Liu, Y., 2015. A fast estimation of distribution algorithm for dynamic fuzzy flexible job-shop scheduling problem. *Computers & Industrial Engineering* 87, 193–201. https://doi.org/10.1016/j.cie.2015.04.029

Mascis, A., Pacciarelli, D., 2002. Job-shop scheduling with blocking and no-wait constraints. *European Journal of Operational Research* 143, 498–517. https://doi.org/10.1016/S0377-2217(01)00338-1

Mohan, J., Lanka, K., Rao, A.N., 2019. A review of dynamic job shop scheduling techniques. *Procedia Manufacturing, Digital Manufacturing Transforming Industry Towards Sustainable Growth* 30, 34–39. https://doi.org/10.1016/j.promfg.2019.02.006

Nowicki, E., Smutnicki, C., 1996. A fast taboo search algorithm for the job shop problem. *Management Science* 42, 797–813. https://doi.org/10.1287/mnsc.42.6.797

Pezzella, F., Morganti, G., Ciaschetti, G., 2008. A genetic algorithm for the flexible job-shop scheduling problem. *Computers & Operations Research, Part Special Issue: Search-based Software Engineering* 35, 3202–3212. https://doi.org/10.1016/j.cor.2007.02.014

Pinedo, M., 2005. *Planning and Scheduling in Manufacturing and Services*, Springer Series in Operations Research. Springer and New York.

Pinedo, M.L., 2012. *Scheduling: Theory, Algorithms, and Systems*, 4th ed. Springer-Verlag, New York. https://doi.org/10.1007/978-1-4614-2361-4

Qiu, X., Lau, H.Y.K., 2014. An AIS-based hybrid algorithm for static job shop scheduling problem. *Journal of Intelligent Manufacturing* 25, 489–503. https://doi.org/10.1007/s10845-012-0701-2

Ramasesh, R., 1990. Dynamic job shop scheduling: A survey of simulation research. *Omega* 18, 43–57.

Ruiz, R., Stützle, T., 2007. A simple and effective iterated greedy algorithm for the permutation flowshop scheduling problem. *European Journal of Operational Research* 177, 2033–2049. https://doi.org/10.1016/j.ejor.2005.12.009

Ruiz, R., Vázquez-Rodríguez, J.A., 2010. The hybrid flow shop scheduling problem. *European Journal of Operational Research* 205, 1–18. https://doi.org/10.1016/j.ejor.2009.09.024

Salveson, M.E., 1952. On a quantitative method in production planning and scheduling. *Econometrica* 20, 554–590. https://doi.org/10.2307/1907643

Serafini, P., Ukovich, W., 1989. A mathematical model for periodic scheduling problems. *SIAM Journal on Discrete Mathematics* 2, 550–581. https://doi.org/10.1137/0402049

Shahzad, A., Mebarki, N., 2016. Learning dispatching rules for scheduling: A synergistic view comprising decision trees, *Tabu Search and Simulation. Computers* 5, 3. https://doi.org/10.3390/computers5010003

Srsen, S., Mernik, M., 2021. A JSSP solution for production planning optimization combining industrial engineering and evolutionary algorithms. *ComSIS* 18, 349–378. https://doi.org/10.2298/CSIS201009058S

Tavakkoli-Moghaddam, R., Jolai, F., Vaziri, F., Ahmed, P.K., Azaron, A., 2005. A hybrid method for solving stochastic job shop scheduling problems. *Applied Mathematics and Computation* 170, 185–206. https://doi.org/10.1016/j.amc.2004.11.036

van Eck, N.J., Waltman, L., 2010. Software survey: VOSviewer, a computer program for bibliometric mapping. *Scientometrics*, 84(2), 523–538.

van Laarhoven, P.J.M., Aarts, E.H.L., Lenstra, J.K., 1992. Job shop scheduling by simulated annealing. *Operations Research* 40, 113–125. https://doi.org/10.1287/opre.40.1.113

Xia, W., Wu, Z., 2005. An effective hybrid optimization approach for multi-objective flexible job-shop scheduling problems. *Computers & Industrial Engineering* 48, 409–425. https://doi.org/10.1016/j.cie.2005.01.018

Yun, Y.S., 2002. Genetic algorithm with fuzzy logic controller for preemptive and non-preemptive job-shop scheduling problems. *Computers & Industrial Engineering* 43, 623–644. https://doi.org/10.1016/S0360-8352(02)00130-4

4 Due-Date Assignment

4.1 INTRODUCTION

In daily life, we constantly have to determine due dates or comply with the due dates set for us and consider them. Although classically it is desired not to miss the due date, according to the JIT philosophy, early completion is also not desired. Therefore, in recent studies, tardiness and early completion are both penalized.

In case of tardiness, the customer is not satisfied; the company suffers from loss of prestige, price reductions, and even loss of customers. In case of early completion, deterioration, stock-keeping costs, and storage space requirements may occur depending on the condition of the products. Demir et al. (2021) have penalized tardiness and earliness, as well as the time length of due dates. No customer will be satisfied with the long due date. Since long due dates can lead to price reductions or loss of customer satisfaction, it is best to give the due date as close as possible. However, an early due date is not always possible or logical. Giving the customer an early and undeliverable due date does not solve the problem. It leaves us in a more difficult situation, and our prestige is damaged in the eyes of the customer. For this reason, we should never give promises we can never keep, for example, giving an unreasonably too early due date to the customers. Demir et al. (2021) weighted customers according to their importance level and suggested giving them reasonably early due dates for important customers and relatively longer due dates for less important ones.

In the literature, a penalty function is mostly used to penalize the sum of early completion and tardiness. On the other hand, Demir et al. (2021) penalized the due-date length together with early completion and tardiness. In addition, the early completion, tardiness, and due date penalties were held more for the important client and less for the less important client. An attempt has been made to give important customers a reasonable early due date and some reasonably distant due dates for less important customers. Thus, the total weighted early completion, tardiness, and due-date penalty costs were tried to be minimized. Demir and his friends have penalized early completion, tardiness, and the length of the due date in many studies they have done in the last ten years (Demir and Erden, 2020; Demir et al., 2021).

4.2 REVIEW OF THE LITERATURE

The problem of scheduling with due-date assignment has been a popular research topic in recent decades. New studies and new approaches have emerged in this field at every stage. Classically, while the due dates were decided externally, they later started to be determined internally.

Scheduling problems can be classified as static vs. dynamic, deterministic vs. stochastic, single product vs. multi-product, single period vs. multi-period, single machine vs. multi-process facilities, and theory vs. practice (Eilon, 1979). Sen and

Gupta (1984) conducted one of the earliest literature reviews on static scheduling, including due-date assignment. They reviewed many studies on this subject up to that time and gave more than a hundred references. They scanned the literature studies for single-machine problems, multi-machine problems, especially two-machine problems, m-parallel machine, and m-flow machine problems. In the studies carried out until then, they stated that the due dates were not integrated with the scheduling and that this issue should be studied. It has been noted that the due dates are not determined as decision variables together with the scheduling. They can be determined, for example, by the marketing department or due to the assembly schedule. It has been stated that organizational performance will increase if they are considered and determined together as decision variables.

Giving reasonable due dates to customers is even more critical in today's competitive environment. There are two essential aspects of the delivery date performance. These are delivery reliability and delivery speed (Hill, 1991). Delivery reliability is the success of meeting the promised delivery date. On the other hand, delivery speed is the success of delivering jobs to customers quickly. Because no customer is willing to accept long due dates, it leaves the company in a difficult situation in this competitive environment with a race against time. Giving a short delivery date will never be a solution if we can't keep our promise; it will even put the company in a more difficult situation. Unkept promises damage the company's prestige significantly. Companies that make fast and reliable deliveries will be especially advantageous in the competitive world. Traditionally, the stock was kept to meet the due dates, and only tardiness was penalized. However, holding stock to meet the delivery date has ceased to be desirable. According to the just-in-time philosophy, earliness and unnecessary stocks are not desirable as well. Therefore, according to this latest situation, we should give time as close as we can keep our promise and be reasonable enough to fulfill our commitments (Philipoom, 2000).

There are assumptions of the common due date and different due dates for delivery dates. In the former, it is the due date used for all parts that must be ready for final assembly by a predetermined date. Again, for the first type, it is used in an MRP system when deciding on the due dates of the lower-level components, which are calculated according to the due date of the main products. The second assumption is valid in a workshop where the due dates of the jobs are determined jointly with the customer and the supplier or in an MTO production system. Another application area of the second assumption is within the JIT production systems that produce and deliver finished products just in time (Zhu and Heady, 2000).

In the previous studies, only delay was penalized, but early completion also became undesirable as JIT-type production spread over. Along with delay, early completion began to be penalized. At this stage, it was discussed how early completion and tardiness would be penalized. In the literature, linear and nonlinear penalty functions were considered for punishment. Then it was thought that the jobs are different and the punishment functions should be different for different jobs. For example, it was thought that deviation from the due date for large jobs was extremely undesirable. Similarly, it was thought that similar penalty functions are not suitable for early completion and delay, and different penalty functions should be used instead. In addition, since the delay is more undesirable, it was thought that the penalty function for tardiness should be a function that is both different and includes more penalties compared to earliness (Zhu and Heady, 2000).

Bank and Werner (2001) dealt with the scheduling problem of n jobs on m unrelated parallel machines. Each job has a ready time greater than 0 and processing times on each machine, and a common due date will be given for all jobs. This study aims to distribute the jobs to the machines so that the weighted sum of linear early completion and delay penalties is minimal.

Biskup and Jahnke (2001) studied the problem of scheduling jobs with a common due date on a single machine and used two different penalty functions. In the first penalty function, early completion and delay penalties were tried to be minimized, and in the second function, the number of delayed jobs was tried to be minimized. As a novelty in the literature, it is assumed that the processing times can be controlled and reduced by the same proportional amount. Mosheiov (2001a) studied scheduling problems with learning effects in one study and discussed assigning common due dates on identical parallel machines in another study (Mosheiov, 2001b).

In previous studies, the problem of assigning due dates was studied, and, later, due windows began to work widely instead of due dates (Chen and Lee, 2002). Instead of specifying a point on the time axis in the due windows, the most appropriate time interval was tried to be determined. While no penalty is applied to the jobs completed within the specified time period, early completion and delay penalties are applied to the completions before and after the window, depending on the situation. Scheduling and determining the due window are explained in detail in Chapter 8.

Gordon et al. (2002a) conducted a literature survey on a single machine and parallel machine scheduling with the due-date assignment and discussed in detail the problem of assigning a common due date in the deterministic case. Gordon et al. (2002b) studied the problem of static deterministic scheduling with the due-date assignment. They analyzed SLK (Common slack due dates), TWK (Total-work-content), PPW (Processing-plus-wait), and NOP (Number-of-operations) due-date assignment rules. They also studied the situation where the due-dates are decided based on the position of the job in the schedule. They conducted the study mainly as a review of the studies from the last ten years, when the date of the study was taken as reference.

Many other studies address single-machine scheduling and due-date assignment problems, such as Shabtay and Steiner (2006) and Biskup and Herrmann (2008), and more recent ones such as Xiong et al. (2018) and Zhao et al. (2018). Parallel machine scheduling with due-date determination studies is also frequently encountered in the literature (Toksari and Güner, 2010; Droubouchevitch and Sidney, 2012). If we include parallel machine scheduling with due window assignment in our review, the following studies can be listed: Chen and Lee (2002); Janiak et al. (2013). The subsequent studies can be examples of scheduling with due-date determination studies in a multi-machine environment (Vinod and Sridharan, 2011; Erden et al., 2019; Demir and Phanden, 2020).

4.3   DUE-DATE ASSIGNMENT APPROACHES

Classically, the due date was usually given exogenously (usually by customers or marketing staff as independent of the production environment). Due dates given independently of production would be unrealistic. If early due dates could not be met, the company's reputation would be damaged, and customers who were not delivered on the promised date would not be satisfied at all. The company sometimes punishes itself for late delivery and reduces prices (Hill, 1989; Keskinocak and Tayur, 2004). In some cases, they had to pay fines imposed by the Federal Trade Commission (FTC) for late delivery (Enos, 2000; Keskinocak and Tayur, 2004). Or, giving an unnecessarily late due date would have no positive effect on the company and the customer. In some studies, in the literature, because due-date lengths are penalized, unnecessarily long due dates lead to customer dissatisfaction or damage to the image of the company and even loss of customers. Sometimes the customer can accept a long due date only in return of a price reduction.

The supply chain paradigm is shifting from mass production to mass customization in a highly competitive global business environment. In such a transformation environment, companies need intelligent decision support tools (Chen et al., 2001). This transformation also shifts from make-to-stock (MTS) to make-to-order (MTO) production. In addition, JIT and lean production have become common as production management systems. According to these new concepts, not only tardiness but also early completion is not desired as classically. All these developments have increased the importance of the due dates given endogenously, taking into account the production environment's situation. Instead of the due dates determined externally in the past, internal due dates optimized according to the firm's conditions become more desirable.

Today, the situation in the MTO production environment and job shop-type productions is between these two extremes. The sales personnel mediate the bargaining between the production and the customer, and the due date is determined after the bargaining (Lawrence, 1994). In addition, these due dates decided with the customer can be the subject of renegotiation later. In renegotiation, a more suitable date is tried to be determined for the firm and the customer. Suppose the firm wants to allocate more capacity to the urgent business of an important customer. In that case, it can make room for important customer orders by negotiating the due date with less important customers (Park et al., 2010).

Another necessary research and application subject is the company's and customer's delivery and price bargaining. It is tried to be agreed on the price according to the given delivery date. Thus, the company does not lose the customer and does not make a promise that it cannot keep, or the service quality is reflected in the price, and the company gains from this situation. The customer either receives quality service or brings the product cheaply in return for waiting. Thus, it has been seen that delivery and price bargaining will be beneficial to the company and the customer (Moodie and Bobrowski, 1999).

Another issue to be considered when assigning a due date is the issue of accepting or rejecting an order. If the firm accepts every order, it may have made a promise that it cannot keep, and if it rejects the orders too often, this may cause resentment in the customer and strengthen the competitors in the competitive world. In this case, instead of accepting every order and giving a due date, the company may reject the least objectionable orders and shorten the due date for other important customers and orders, making its production possible and increasing the delivery quality.

4.4   DUE-DATE ASSIGNMENT RULES

Some system features distinguish the due-date assignment methods from each other. These features are offline vs. online, single vs. multiple servers, stochastic vs. deterministic processing times, common vs. distinct due dates, setup times/costs, and others. All jobs can be ready in the workshop or come to the workshop according to a distribution, like exponential inter-arrival time. There may be one or more similar or dissimilar machines in the workshop. Process times can be deterministic or random. Jobs on the shop floor can be given a common due date, and in many cases, each job can be given a particular due date. In some problems, it may be necessary to consider setup times, which can affect due dates (Keskinocak and Tayur, 2004).

Some due date assignments, such as the Constant (CON) and Random (RDM) methods, are external. In this case, the due dates are determined as soon as the arrival of jobs, and these due dates are fixed and determined independently of the jobs' characteristics and the shop floor's characteristics (Cheng and Gupta, 1989). Before moving on to the due-date rules, the notations are given in Table 4.1.

(Weighted) Constant Flow Allowance ((W)CON): When this rule is used, a constant flow allowance is given to all the jobs coming to the shop floor, and the due date of the incoming jobs using the same constant is given externally regardless of the jobs and shop floor characteristics. In weighted cases, important customers are given closer due dates, meaning the constant flow allowance will be shorter. If all the jobs are ready on the shop floor at time zero, this method turns into a common due-date assignment method. $d_j = q$, and this rule is called the common due-date assignment model, and all jobs have a due date of D.

This rule is used for jobs that are pending for assembly and need to be completed at the same time, or when jobs that create orders for the same customer need

TABLE 4.1
Notations Used in Due-date Assignment

d_j	: (Weighted) Due-date of job j	C_j	: Completion time of job j
a_j/r_j	: Arrival/Release time of job j	l_j	$= d_j - a_j$ Lead Time of job j
p_j	: Processing time of job j	F_j	$= C_j - a_j$ Flow Time of job j
n_j	: Number of operations of job j	E_j	: Earliness of job j $E_j = Max\{d_j - C_j, 0\}$
P_{avg}	: Mean processing time of all jobs waiting	T_j	: Tardiness of job j $T_j = Max\{C_j - d_j, 0\}$
w_j	: Weight of job j	L_j	: Lateness of job j $L_j = C_j - d_j$
k_x	: A multiplier such as 1, 2, 3, 5, 10 is used in (W) TWK, (W) NOP, and (W)PPW	e_j	: Random flow allowance representing the customer-specified due-date for job j
Z_x, Z_y	: Changes according to the weights of each customer. If the weight is big, then Z_x and Z_y values become appropriately small to give closer due-date	q_x	: Constant slack used in (W)CON, (W)SLK, and (W)PPW such as ($P_{avg}/2$), (P_{avg}), or ($3*P_{avg}/2$)
W_j	$= d_j - a_j - p_j$ Waiting time of job j		

to be completed at the same time, or where there are jobs that need to be completed at the same time due to some constraints.

Random Due-Date Assignment (RDM): In this rule, a random flow allowance is assigned when assigning a due date for any job. Random flow allowance refers to the random external due dates notified to the shop floor by the customers and that we must produce beyond the control of the shop floor.

Some due-date determination methods are internal. In this case, the shop floor decides on the due date under its own situation. Since there is no obligation to comply with the due date dictated from the outside, the due dates can be optimized in the company's interest and in accordance with the situation on the shop floor. This is a significant opportunity and a great gain for the firm.

(W)SLK ((Weighted) Slack) Due-date Determination Method: In this method, an equal slack time is added to each job that comes to the shop floor in addition to the processing time, and thus the due date time of that job is determined. Here, it is tried to determine the most suitable slack period for the firm's maximum benefit and the situation of the shop floor. The weighted rule gives the due date relatively earlier for important customers.

(W)TWK ((Weighted) Total Work) Due-Date Determination Method: The due dates are determined depending on the total work content. The processing times are multiplied by a common coefficient for the incoming jobs so that the due dates of each job are determined. Here, it is tried to determine the coefficient "k" that will be most beneficial to the enterprise and most suitable for the conditions of the shop floor. The weighted rule gives the due date relatively earlier for important customers.

(W)PPW ((Weighted) Processing plus wait) Due-Date Determination Method: This method determines the due date of each job based on the total job content and the fixed slack time to be used. The total work content is multiplied by an optimized coefficient as in TWK, and an optimized fixed slack time is added as in the SLK. Again, in weighted cases, important customers get early due dates.

(W)NOP ((Weighted) number of operations) Due-Date Determination Method: This method uses each job's number of operations (n_j). The number of operations of each job is multiplied by an optimized coefficient to assign due dates. Here, the coefficient "k" is optimized most appropriately to the enterprise's interests and the shop floor's specific situation. Major due-date assignment rules for weighted and unweighted cases are summarized in Table 4.2.

There are other rules available that are used other than the common rules described here. The "TWK+NOP" rule is a hybrid of the TWK and NOP rules. The "JIQ" rule sets a due date, considering the jobs waiting to be processed before job j (the number of jobs waiting in the queue). The "JIS" rule takes into account the jobs waiting in the system (number of jobs in the system) at the ready time of job j. "WIQ" considers the total processing times of jobs waiting in the queue before job j. "WINS" takes into account the sum of the processing times of all jobs in the system. Apart from these rules, many other ones are used in the literature (Keskinocak and Tayur, 2004).

TABLE 4.2
Due-date Assignment Rules

Rule	Explanation	Formula
WCON/CON	(Weighted) constant due-date	$Dj = r_j + (Z_x)q$
WSLK/SLK	(Weighted) slack	$Dj = r_j + p_j + (Z_x)q_x$
WTWK/TWK	(Weighted) total work content	$Dj = r_j + (Z_x)k_xp_j$
WNOP/NOP	(Weighted) number of operations	$Dj = r_j + (Z_x)k_xNOP$
WPPW/PPW	(Weighted) processing-time-plus-wait	$Dj = r_j + (Z_x)k_xp_j + (Z_y)q_y$
RDM	Random-allowance due-dates	$Dj = r_j + e_j$

Z_x, Z_y parameters assume inversely proportional values to the customer weights.

4.5 WEIGHTING THE CUSTOMERS

In some studies, due dates are punished alongside earliness and tardiness in the literature. No reasonable customer is satisfied with a late due date. Sometimes, a late due date may be accepted with a price reduction, and long due dates increase customer dissatisfaction. For these reasons, it would be appropriate to punish the due dates and avoid unnecessarily long due dates. In the literature, penalty functions are mostly calculated independently of customer weights. But the dissatisfaction of the important customer will be a bigger problem for the firm, and it would be appropriate to calculate the higher penalty costs for the important customers. Therefore, customer weights must be taken into account for high penalty costs.

In addition, if the penalty cost is high for important customers, then the improvement in the situation of important customers will significantly improve the penalty function. In their studies, Demir et al. (2021) suggested giving weighted due dates to important customers, that is, to give close due dates to important customers and to apply weighted scheduling for these important works, that is, early scheduling. Thus, since the gain in important customers can be much more than the losses in less important customers, this will make a significant contribution to the firm (Erden et al., 2019; Demir and Phanden, 2020; Demir et al., 2021).

4.6 CONCLUSION

Classically, only tardiness was penalized. Over time, as a result of the spread of the JIT philosophy, early completions also became undesirable. In addition, internal due dates determined by the shop floor and production started to be adopted instead of the due dates determined externally independent from the shop floor and the production. On the internally determined due dates, the company began to determine the most suitable due dates for the situation of the shop floor and its benefit.

Some studies show that due dates are penalized in addition to tardiness and earliness. In these studies, the due date length is also penalized. Although tardiness, earliness, and due date are penalized linearly in many studies, there are also cases where tardiness is penalized with a quadratic function. Although both tardiness and earliness are not desired, penalty costs are considered higher in case of tardiness. Significant customer dissatisfaction and related costs occur in tardiness and long due dates. The company loses prestige, sometimes discounts can be made on product prices, and worst of all, customer losses may occur in the future. On the early due dates, the customer can be satisfied and even be willing to pay a little more for the nearer dates, and the prestige of the company increases positively.

However, the close due dates given without being informed of the shop floor conditions cannot be met, which may lead to worse results. When some customers are given an early due date, the service performance of other customers can be badly affected. In some studies, customer weights were used, and it was thought it would be right first to gain the satisfaction of important customers. It can bring significant gains for the firm when the benefit for important customers is significantly greater than the loss for less important customers. Early completion is also very undesirable in JIT and Lean Manufacturing philosophies today. For unnecessarily early completed works, costs such as keeping stock, occupying space, allocating the capital to the product early, and in some cases, spoilage cost of the products occur.

As a result, penalty functions with customer weights, tardiness is penalized more, due dates are penalized, and earliness is not desired also. Internal due-date assignments, in which the most appropriate due dates are determined for the conditions of the shop floor and the benefit of the company, have been a popular study topic recently.

REFERENCES

Bank, J., & Werner, F. (2001). Heuristic algorithms for unrelated parallel machine scheduling with a common due date, release dates, and linear earliness and tardiness penalties. *Mathematical & Computer Modelling*, 33(4), 363–383.

Biskup, D., & Herrmann, J. (2008). Single-machine scheduling against due dates with past-sequence-dependent setup times. *European Journal of Operational Research*, 191(2), 587–592.

Biskup, D., & Jahnke, H. (2001). Common due date assignment for scheduling on a single machine with jointly reducible processing times. *International Journal of Production Economics*, 69(3), 317–322.

Chen, C.-Y., Zhao, Z.-Y., & Ball, M. O. (2001). Quantity and due date quoting available to promise. *Information Systems Frontiers*, 3(4), 477–488.

Chen, Z.-L., & Lee, C.-Y. (2002). Parallel machine scheduling with a common due window. *European Journal of Operational Research*, 136(3), 512–527.

Cheng, T. C. E., & Gupta, M. C. (1989). Survey of scheduling research involving due date determination decisions. *European Journal of Operational Research*, 38(2), 156–166. https://doi.org/10.1016/0377-2217(89)90100-8

Demir, H. I., & Erden, C. (2020). Dynamic integrated process planning, scheduling and due-date assignment using ant colony optimization. *Computers & Industrial Engineering*, 149, 106799. https://doi.org/10.1016/j.cie.2020.106799

Demir, H. I., & Phanden, R. K. (2020). Due-date agreement in integrated process planning and scheduling environment using common meta-heuristics. In *Integration of Process Planning and Scheduling: Approaches and Algorithms*. Ed: Phanden, R. K., Jain, A. & Davim, J.P. (pp. 161–182). CRC Press/Taylor & Francis Group, Boca Raton.

Demir, H. I., Phanden, R., Kökçam, A., Erkayman, B., & Erden, C. (2021). Hybrid evolutionary strategy and simulated annealing algorithms for integrated process planning, scheduling and due-date assignment problem. *Academic Platform Journal of Engineering and Science*, 9(1), 86–91. https://doi.org/10.21541/apjes.764150

Droubouchevitch, I. G., & Sidney, J. B. (2012). Minimization of earliness, tardiness and due date penalties on uniform parallel machines with identical jobs. *Computers & Operations Research*, 39(9), 1919–1926.

Eilon, S. (1979). Production scheduling. In *OR '78* (Ed. Hayley, K. B., pp. 237–266). North Holland, Amsterdam, Amsterdam.

Enos, L. (2000). Report: Holiday e-sales to double. *E-Commerce Times*, September 6. http://www.ecommercetimes.com/perllstory/4202.htrnl

Erden, C., Demir, H. I., & Kökçam, A. H. (2019). Solving integrated process planning, dynamic scheduling, and due date assignment using metaheuristic algorithms. *Mathematical Problems in Engineering*, 2019, 1–19. https://doi.org/10.1155/2019/1572614

Gordon, V., Proth, J.-M., & Chu, C. (2002a). A survey of the state-of-the-art of common due date assignment and scheduling research. *European Journal of Operational Research*, 139(1), 1–25. https://doi.org/10.1016/S0377-2217(01)00181-3

Gordon, V. S., Proth, J.-M., & Chu, C. (2002b). Due date assignment and scheduling: SLK, TWK and other due date assignment models. *Production Planning & Control*, 13(2), 117–132. https://doi.org/10.1080/09537280110069621

Hill, N. (1989). Delivery on the dot- or a refund! *Industrial Marketing Digest*, 14(2), 43–50.

Hill, T. (1991). *Production/Operations Management, Text and Cases*. New York: Prentice-Hall.

Janiak, A., Janiak, W., Kovalyov, M. Y., Kozan, E., & Pesch, E. (2013). Parallel machine scheduling and common due window assignment with job independent earliness and tardiness costs. *Information Sciences*, 224, 109–117. https://doi.org/10.1016/j.ins.2012.10.024

Keskinocak, P., & Tayur, S. (2004). Due date management policies. In *Handbook of Quantitative Supply Chain Analysis: Modeling in the E-Business Era* (pp. 485–554). New York, Springer.

Lawrence, S. R. (1994). Negotiating due-dates between customers and producers. *International Journal of Production Economics*, 37(1), 127–138. https://doi.org/10.1016/0925-5273(94)90013-2

Moodie, D. R., & Bobrowski, P. M. (1999). Due date demand management: Negotiating the trade-off between price and delivery. *International Journal of Production Research*, 37(5), 997–1021.

Mosheiov, G. (2001a). Scheduling problems with a learning effect. *European Journal of Operational Research*, 132(3), 687–693.

Mosheiov, G. (2001b). A common due-date assignment problem on parallel identical machines. *Computers & Operations Research*, 28(8), 719–732. https://doi.org/10.1016/S0305-0548(99)00127-6

Park, M., Lee, D., Shin, K., & Park, J. (2010). Business integration model with due-date renegotiations. *Industrial Management & Data Systems*, 110(3), 415–432.

Philipoom, P. R. (2000). The choice of dispatching rules in a shop using internally set due-dates with quoted leadtime and tardiness costs. *International Journal of Production Research*, 38(7), 1641–1655.

Sen, T., & Gupta, S. K. (1984). A state-of-art survey of static scheduling research involving due dates. *Omega*, 12(1), 63–76. https://doi.org/10.1016/0305-0483(84)90011-2

Shabtay, D., & Steiner, G. (2006). Two due date assignment problems in scheduling a single machine. *Operations Research Letters*, 34(6), 683–691. https://doi.org/10.1016/j.orl.2005.10.009

Toksari, M. D., & Güner, E. (2010). The common due-date early/tardy scheduling problem on a parallel machine under the effects of time-dependent learning and linear and nonlinear deterioration. *Expert Systems with Applications*, 37(1), 92–112.

Vinod, V., & Sridharan, R. (2011). Simulation modeling and analysis of due-date assignment methods and scheduling decision rules in a dynamic job shop production system. *International Journal of Production Economics*, 129(1), 127–146. https://doi.org/10.1016/j.ijpe.2010.08.017

Xiong, X., Wang, D., Edwin Cheng, T. C., Wu, C.-C., & Yin, Y. (2018). Single-machine scheduling and common due date assignment with potential machine disruption. *International Journal of Production Research*, 56(3), 1345–1360. https://doi.org/10.1080/00207543.2017.1346317

Zhao, C., Hsu, C.-J., Lin, W.-C., Liu, S.-C., & Yu, P.-W. (2018). Due date assignment and scheduling with time and positional dependent effects. *Journal of Information and Optimization Sciences*, 39(8), 1613–1626. https://doi.org/10.1080/02522667.2017.1367515

Zhu, Z., & Heady, R. B. (2000). Minimizing the sum of earliness/tardiness in multi-machine scheduling: A mixed integer programming approach. *Computers & Industrial Engineering*, 38(2), 297–305. https://doi.org/10.1016/S0360-8352(00)00048-6

5 Integrated Process Planning and Scheduling

5.1 INTRODUCTION: BACKGROUND AND DRIVING FORCES

Why are integrated function studies needed in manufacturing systems? The simple answer is to achieve optimization goals and master production plans in a supply chain more effectively. If one process can be input into another, it is thought that it would be helpful to handle these two processes together. Process planning and scheduling are critical functions in manufacturing systems. Both process planning and scheduling functions are at a point that directly affects efficiency and the capacity usage of manufacturing resources (Yang et al., 2001). The outputs obtained in the process planning are used as inputs during the scheduling phase. As mentioned, studies are conducted in the process planning phase, in which processes should be completed to produce a product (Chang et al., 1985). Production resources are assigned to production operations in the scheduling phase, and schedules are generated according to the weight of jobs and optimization criteria (Li & McMahon, 2007a). Therefore, scheduling systems are optimization problems in many areas, such as personnel scheduling, CPU scheduling or project management, and manufacturing systems (Pinedo, 2012). Scheduling systems optimize many objectives, such as makespan, mean flow time, balancing the level of machine loads, maximizing the number of products, more reliable date dates for each job, and minimum manufacturing costs (Mönch et al., 2011).

Traditionally, scheduling and process planning are performed separately, in which the scheduling process is initiated with a proper process plan for each part or job. This restricts scheduling by using only one machine sequence in each operation. Therefore, the possible resources used in the scheduling can be ignored. In addition, although process plans are often based on technological needs, scheduling is often related to timing and allocation. Combining process planning and scheduling functions can solve many problems in a manufacturing system to increase optimality and provide optimal use of production resources (Huang et al., 1995). Although the differences between process planning and scheduling are shown artificially, as described by Rippin (1993), these two functions can be separated by considering the length of their time and related decisions. Wilhelm and Shin (1985) discuss the efficiency of alternative operations. The researchers then studied integrated process plans and schedules that increase productivity in production.

This section presents the integrated process planning and scheduling as integrated functions, as previously discussed individually. Firstly, it will review the literature on IPPS. Secondly, it proposes a conceptual framework and mathematical model for IPPS. Finally, it addresses some optimization algorithms developed to solve IPPS, such as genetic algorithms, particle swarm optimization, and ant colony algorithms.

DOI: 10.1201/9781003215295-5

5.2   REVIEW OF THE LITERATURE

As mentioned earlier, many studies have been conducted on integrated process planning and scheduling (IPPS) over the past decade. IPPS was first defined in 1985 by Chryssolouris et al. (1985) as a decision-making approach. According to a recent definition, Guo et al. (2009) defined the IPPS problem as follows.

> Given a set of n parts which are to be processed on m machines with operations including alternative manufacturing resources (machines and tools), select suitable manufacturing resources and sequence the operations to determine a schedule in which the precedence constraints among operations can be satisfied, and the corresponding objectives can be achieved.

To solve the IPPS problem, researchers first created a decision matrix to select the alternatives. In 1988, Sundaram and Fu (1988) developed a scheduling approach to minimize cycle time. They developed an integrated computer-aided process planning and scheduling automation system based on the group technology. The developed system has been published as Integrated Computer-Aided Process Planning and Scheduling (ICAPPS) (Sundaram & Fu, 1988). Additionally, Srihari and Greene (1990) produced the first sample solution, CAPP 1990.

Integrating process planning and scheduling functions has emerged from the result that working with alternative plans provides higher efficiency in production environments. Wilhelm and Shin (1985) discuss the efficiency of alternative operations. The researchers then studied integrated process plans and schedules to increase the efficiency of production plans. The primary purpose of the integration model is to ensure that process planning and scheduling functions work together. In addition, here are the following advantages and aims (Zhang & Mallur, 1994):

- Create process plans that reflect shop floor conditions and respond to the dynamic structure of the production system.
- To prepare a basis for Just-in-Time Production (JIT). This approach allows the preparation of process plans as needed.
- Achieving the best process plans by considering the objectives of both process planning and scheduling functions.
- To increase the efficiency of the production system by improving machine use and reducing processing costs and time
- Developing integration between related functions brings production closer to a fully integrated system.
- It is easier for all production functions to access process planning and scheduling information for improved information storage.

The best way to integrate process planning and scheduling is to handle the two functions as a single optimization problem. Process planning and scheduling functions are solely NP-hard problems, and it is difficult to find optimal results and times to solve when they are integrated (Khoshnevis & Chen, 1991; Kis, 2003). However, many studies have been conducted on integrating these two functions. This is because IPPS contributes to increasing production efficiency and capacity utilization rates. IPPS approaches were examined with the study in many surveys that can be given (Gen et al., 2009; Li et al., 2010; Phanden et al., 2011; Tan & Khoshnevis, 2000).

It is seen in the literature that many IPPS studies are solved with metaheuristics. Among metaheuristics, GA is one of the most widely used algorithms (Amin-Naseri & Afshari, 2012; Bierwirth & Mattfeld, 1999; Chan et al., 2006; Ip et al., 2000; Moon et al., 2002; Morad & Zalzala, 1999). Morad and Zalzala (1999) optimized the IPPS problem using GA from these studies. Qiao and Lv (2014) solved an integrated process planning and scheduling/rescheduling problem by improving the GA algorithm. PSO, a well-known algorithm, has been used for IPPS. Zhao et al. (2010) solved the IPPS problem using a hybrid particle swarm optimization (PSO) model and fuzzy logic in their study. Li et al. (2013) attempted to solve the IPPS problem using an effective updated PSO algorithm. In addition to these studies, examples of other algorithms are as follows: evolutionary algorithms (Kim et al., 2003a; Li et al., 2010; Moon & Seo, 2005) and Game Theory (Li et al., 2012; Phanden et al., 2013), Honey Bee Optimization (Li et al., 2015; Yuan et al., 2015), simulated annealing (Lee & Kim, 2001; Li & McMahon, 2007b; Palmer, 1996), hybrid algorithms (Ausaf et al., 2015; Yu et al., 2015), ant colony optimization (Leung et al., 2010). Some studies have generated different perspectives on IPPS. Kim et al. (2003b) studied solutions to the IPPS problem using operational, sorting, and process flexibilities. Yan et al. (2003) addressed the IPPS problem to minimize tardiness.

5.3 MODELING IPPS

Scheduling problems are focused on a limited number of jobs (n) and a limited number of machines (m). Typically, subscript j denotes the number of jobs, subscript i denotes the number of machines, and k denotes the number of routes. Table 5.1 lists the notations used in the problem (Pinedo, 2012).

TABLE 5.1
Notations

Symbol	Explanation
G	Routes for each job
O	Operations
DD_j	Due date of job j
w_j	Weight of job j
p_j	Processing time of job j
p_{ij}	Processing time of job j at machine i
p_{av}	Average processing time
c_j	Completion time of job j
L_j	Lateness time of job j
T_j	Tardiness time of job j
E_j	Earliness time of job j
α	Earliness penalty coefficient, $\alpha > 0$,
β	Tardiness penalty coefficient, $\beta > 0$,
$f_{\min}$	Objective function

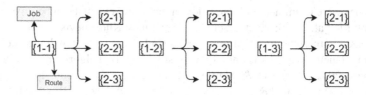

FIGURE 5.1 Numbers of alternative process plans.

The most crucial stage in IPPS is the alternative process plan. Jobs that arrive on the shop floor can take more than one route. An extremely high number of combinations must be attempted to determine the route from these alternative process plans, which should be the input of the scheduling phase. All alternatives, $G_1, \ldots, G_r$ must be evaluated to determine the routes. The formula (k^n) is used to calculate the number of alternative routes. For example, if the number of jobs is $n = 2$ and the number of routes is $k = 3$, there may be $3^2 = 9$ alternative routes (see Figure 5.1) in which {job -route} notation is used. As the number of routes and jobs increases, so does the number of alternative routes (see Figure 5.1).

Three types of flexibility can be mentioned when dealing with process planning and scheduling systems: operational, sequencing, and processing flexibility. Operational flexibility implies that an operation can be performed on alternative machines. With the flexibility of sequencing, it is possible to change the order of operation when necessary. Process flexibility is the possibility of production characteristics similar to alternative operations or other operation sequences (Benjaafar & Ramakrishnan, 1996; Hutchinson & Pflughoeft, 1994; Saygin & Kilic, 1999). Some graphical representations in the literature represent flexibilities, such as Petri net (K. Lee & Jung, 1994), AN/OR, and network display models. In this study, network notation was chosen as it is problem-appropriate, and its representation is straightforward. In addition to the essential representation, the due dates and weights of jobs have been added to the network representation (see Figure 5.2).

5.4 MODELING A SAMPLE NETWORK REPRESENTATION FOR ALTERNATIVE PROCESS PLANS

The objective function of scheduling problems is usually to minimize the completion time or makespan. Some studies have also maximized machine workloads, minimized the average flow time, balanced machine utilization, and minimized job tardiness or manufacturing costs. The lateness of jobs is calculated as follows:

$$Lj = Cj - dj \tag{5.1}$$

If the delay here is positive, the job j is finished late, and if it is negative, the job j is finished early.

$$T_j = \max\left(C_j - d_j, 0\right) = \max\left(L_j, 0\right) \tag{5.2}$$

$$E_j = \max\left(d_j - C_j, 0\right) \tag{5.3}$$

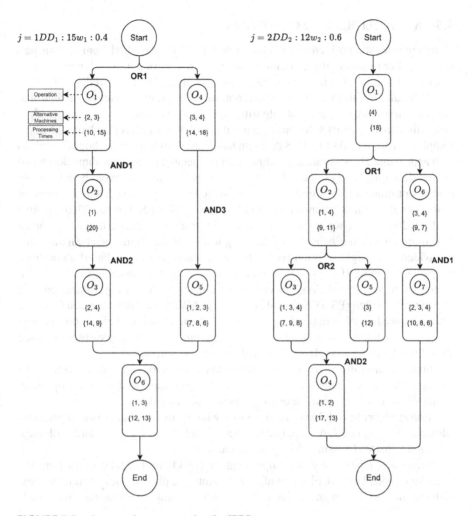

FIGURE 5.2 A network representation for IPPS.

If the job is finished early, the difference between the delivery and completion times will not be negative. The sum of the early and late completion times can be considered an objective function.

$$\sum_{j=1}^{n} E_j + \sum_{j=1}^{n} T_j \tag{5.4}$$

Weighting can be applied for early and late completion to obtain a better objective function.

$$f_{\min} = \sum_{j=1}^{n} \alpha E_j + \sum_{j=1}^{n} \beta T_j \tag{5.5}$$

5.5   METAHEURISTIC SOLUTIONS

Many optimization problems in the literature have been classified from various perspectives. For example, if the solution contains a derivative computation or gradient function, it can be classified as a gradient-based algorithm (Creswell & Creswell, 2017). In addition to this classification, optimization problems can be divided into exact and approximately accurate algorithms. IPPS is challenging to solve with algorithms that provide exact solutions because it is an NP-hard class problem with many complex constraints. As the IPPS problem has a vast search space, it should be solved with approximately accurate algorithms, that is, metaheuristic algorithms developed to provide a sufficiently good solution in a reasonable time and avoid being trapped in local optimum points. Metaheuristics, classified in many ways, can find optimal or near-optimal global solutions in any problem space (Okwu & Tartibu, 2020). Molina et al. (2020) created well-defined categories by evaluating two criteria: the source of inspiration and the behavior of each algorithm. In addition, metaheuristics can be divided into single solution-based (trajectory) and population-based algorithms, as shown in Figure (Lones, 2014). Well-known population-based algorithms, such as genetic algorithms (GAs), ant colony optimization (ACO) algorithms, particle swarm optimization (PSO), artificial bee colony (ABC), and cuckoo search (CS), are nature-inspired metaheuristics (see Figure 5.3). Population-based algorithms change solutions at each iteration as the next generation, whereas single-solution-based algorithms focus on a single solution and improve its solution.

Simulated annealing (SA) mimics metal annealing (Kirkpatrick et al., 1983). The SA algorithm uses a heuristic process to avoid being trapped at the local optimum points. Thus, it differs from the other gradient-based algorithms.

Genetic algorithms (GAs) are the most widely used metaheuristic algorithms (Holland, 1992). The GA is population-based and mimics genetic and biological evolution (Charles Darwin's theory of natural selection).

Dorigo developed ant colony optimization (ACO) in 1992 by mimicking the behavior of social ants (Colorni et al., 1992). Ants use pheromones to communicate with one another. When an ant finds a food source, it drops pheromones on its road.

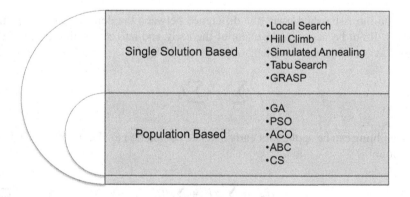

FIGURE 5.3   A taxonomy of metaheuristic optimization algorithms.

Kennedy and Eberhart developed particle swarm optimization (PSO) in 1995 that mimics the swarm behavior observed in nature, such as fish and birds (Eberhart & Kennedy, 1995).

Fred Glover developed a tabu search (TS) in the 1970s; tabu search explicitly uses memory, and the search history is a significant component of the method (Glover & Laguna, 1998).

In the literature, metaheuristic algorithms have been approached from many angles to solve the IPPS problem. Studies in this field are mentioned in Chapter 11. Therefore, short entry-level information regarding the algorithms selected in this section is provided.

5.6 CONCLUSION

This section describes the IPPS problem, which combines the two functions of the integration levels. The IPPS problem can be considered the basis for more extensive integration. Therefore, understanding the IPPS problem and establishing the reasons for its use will also provide benefits for more extensive integration. Today, IPPS problems have been frequently studied and can be solved using heuristic algorithms. As many different heuristic algorithms exist and new algorithms have emerged over time, this has led to numerous studies on the IPPS problem. Publications can be made using each newly derived algorithm. Including such problems in the literature constitutes a new and empty field. However, finding a gap in the literature is challenging. We see that the literature continues with the trial of different algorithms on IPPS. Some of the studies are mentioned in this section, and IPPS modeling is presented.

REFERENCES

Amin-Naseri, M. R., & Afshari, A. J. (2012). A hybrid genetic algorithm for integrated process planning and scheduling problem with precedence constraints. *The International Journal of Advanced Manufacturing Technology*, 59(1–4), 273–287.

Ausaf, M. F., Gao, L., Li, X., & Al Aqel, G. (2015). A priority-based heuristic algorithm (PBHA) for optimizing integrated process planning and scheduling problem. *Cogent Engineering*, 2(1), 1070494.

Benjaafar, S., & Ramakrishnan, R. (1996). Modelling, measurement and evaluation of sequencing flexibility in manufacturing systems. *International Journal of Production Research*, 34(5), 1195–1220.

Bierwirth, C., & Mattfeld, D. C. (1999). Production scheduling and rescheduling with genetic algorithms. *Evolutionary Computation*, 7(1), 1–17.

Chan, F. T., Kumar, V., & Tiwari, M. K. (2006). Optimizing the performance of an integrated process planning and scheduling problem: An AIS-FLC based approach. *2006 IEEE Conference on Cybernetics and Intelligent Systems*, 1–8.

Chang, Y., & Wysk, R.A. (1985). *An Introduction to Automated Process Planning*. Prentice-Hall.

Chryssolouris, G., Chan, S., & Suh, N. P. (1985). An integrated approach to process planning and scheduling. *CIRP Annals*, 34(1), 413–417.

Colorni, A., Dorigo, M., & Maniezzo, V. (1992). An investigation of some properties of an "Ant Algorithm". *PPSN*, 92, 509–520. http://staff.washington.edu/paymana/swarm/colorni92-ppsn.pdf

Creswell, J. W., & Creswell, J. D. (2017). *Research Design: Qualitative, Quantitative, and Mixed Methods Approaches*. Sage Publications.

Eberhart, R., & Kennedy, J. (1995). Particle swarm optimization. *Proceedings of the IEEE International Conference on Neural Networks, 4*, 1942–1948.

Gen, M., Lin, L., & Zhang, H. (2009). Evolutionary techniques for optimization problems in integrated manufacturing system: State-of-the-art-survey. *Computers & Industrial Engineering, 56*(3), 779–808.

Glover, F., & Laguna, M. (1998). Tabu search. In D.-Z. Du & P. M. Pardalos (Eds.), *Handbook of Combinatorial Optimization* (pp. 2093–2229). Springer. https://doi.org/10.1007/978-1-4613-0303-9_3

Guo, Y. W., Li, W. D., Mileham, A. R., & Owen, G. W. (2009). Applications of particle swarm optimisation in integrated process planning and scheduling. *Robotics and Computer-Integrated Manufacturing, 25*(2), 280–288.

Holland, J. H. (1992). Genetic algorithms. *Scientific American, 267*(1), 66–73.

Huang, S. H., Zhang, H.-C., & Smith, M. L. (1995). A progressive approach for the integration of process planning and scheduling. *IIE Transactions, 27*(4), 456–464.

Hutchinson, G. K., & Pflughoeft, K. A. (1994). Flexible process plans: Their value in flexible automation systems. *The International Journal of Production Research, 32*(3), 707–719.

Ip, W. H., Li, Y., Man, K. F., & Tang, K. S. (2000). Multi-product planning and scheduling using genetic algorithm approach. *Computers & Industrial Engineering, 38*(2), 283–296.

Khoshnevis, B., & Chen, Q. M. (1991). Integration of process planning and scheduling functions. *Journal of Intelligent Manufacturing, 2*(3), 165–175.

Kim, Y. K., Park, K., & Ko, J. (2003a). A symbiotic evolutionary algorithm for the integration of process planning and job shop scheduling. *Computers & Operations Research, 30*(8), 1151–1171.

Kim, Y. K., Park, K., & Ko, J. (2003b). A symbiotic evolutionary algorithm for the integration of process planning and job shop scheduling. *Computers & Operations Research, 30*(8), 1151–1171.

Kirkpatrick, S., Gelatt, C. D., & Vecchi, M. P. (1983). Optimization by simulated annealing. *Science, 220*(4598), 671–680.

Kis, T. (2003). Job-shop scheduling with processing alternatives. *European Journal of Operational Research, 151*(2), 307–332. https://doi.org/10.1016/S0377-2217(02)00828-7

Lee, H., & Kim, S.-S. (2001). Integration of process planning and scheduling using simulation based genetic algorithms. *The International Journal of Advanced Manufacturing Technology, 18*(8), 586–590.

Lee, K., & Jung, M. (1994). Petri net application in flexible process planning. *Computers & Industrial Engineering, 27*(1–4), 505–508.

Leung, C. W., Wong, T. N., Mak, K.-L., & Fung, R. Y. (2010). Integrated process planning and scheduling by an agent-based ant colony optimization. *Computers & Industrial Engineering, 59*(1), 166–180.

Li, W. D., & McMahon, C. A. (2007a). A simulated annealing-based optimization approach for integrated process planning and scheduling. *International Journal of Computer Integrated Manufacturing, 20*(1), 80–95.

Li, W. D., & McMahon, C. A. (2007b). A simulated annealing-based optimization approach for integrated process planning and scheduling. *International Journal of Computer Integrated Manufacturing, 20*(1), 80–95.

Li, X., Gao, L., & Li, W. (2012). Application of game theory based hybrid algorithm for multi-objective integrated process planning and scheduling. *Expert Systems with Applications, 39*(1), 288–297.

Li, X., Gao, L., Shao, X., Zhang, C., & Wang, C. (2010). Mathematical modeling and evolutionary algorithm-based approach for integrated process planning and scheduling. *Computers & Operations Research, 37*(4), 656–667.

Li, X., Gao, L., & Wen, X. (2013). Application of an efficient modified particle swarm optimization algorithm for process planning. *The International Journal of Advanced Manufacturing Technology, 67*(5), 1355–1369.

Li, X., Gao, L., Zhang, C., & Shao, X. (2010). A review on integrated process planning and scheduling. *International Journal of Manufacturing Research, 5*(2), 161–180.

Li, X. X., Li, W. D., Cai, X. T., & He, F. Z. (2015). A hybrid optimization approach for sustainable process planning and scheduling. *Integrated Computer-Aided Engineering, 22*(4), 311–326.

Lones, M. A. (2014). Metaheuristics in nature-inspired algorithms. *Proceedings of the Companion Publication of the 2014 Annual Conference on Genetic and Evolutionary Computation*, 1419–1422.

Lv, S., & Qiao, L. (2014). Process planning and scheduling integration with optimal rescheduling strategies. *International Journal of Computer Integrated Manufacturing, 27*(7), 638–655.

Molina, D., Poyatos, J., Ser, J. D., García, S., Hussain, A., & Herrera, F. (2020). Comprehensive taxonomies of nature- and bio-inspired optimization: Inspiration versus algorithmic behavior, critical analysis recommendations. *Cognitive Computation, 12*(5), 897–939. https://doi.org/10.1007/s12559-020-09730-8

Mönch, L., Fowler, J. W., Dauzère-Pérès, S., Mason, S. J., & Rose, O. (2011). A survey of problems, solution techniques, and future challenges in scheduling semiconductor manufacturing operations. *Journal of Scheduling, 14*(6), 583–599. https://doi.org/10.1007/s10951-010-0222-9

Moon, C., Kim, J., & Hur, S. (2002). Integrated process planning and scheduling with minimizing total tardiness in multi-plants supply chain. *Computers & Industrial Engineering, 43*(1–2), 331–349.

Moon, C., & Seo, Y. (2005). Evolutionary algorithm for advanced process planning and scheduling in a multi-plant. *Computers & Industrial Engineering, 48*(2), 311–325.

Morad, N., & Zalzala, A. M. S. (1999). Genetic algorithms in integrated process planning and scheduling. *Journal of Intelligent Manufacturing, 10*(2), 169–179.

Okwu, M. O., & Tartibu, L. K. (2020). *Metaheuristic Optimization: Nature-Inspired Algorithms Swarm and Computational Intelligence, Theory and Applications* (Vol. 927). Springer Nature.

Palmer, G. J. (1996). A simulated annealing approach to integrated production scheduling. *Journal of Intelligent Manufacturing, 7*(3), 163–176.

Phanden, R. K., Jain, A., & Verma, R. (2011). Integration of process planning and scheduling: A state-of-the-art review. *International Journal of Computer Integrated Manufacturing, 24*(6), 517–534.

Phanden, R. K., Jain, A., & Verma, R. (2013). An approach for integration of process planning and scheduling. *International Journal of Computer Integrated Manufacturing, 26*(4), 284–302.

Pinedo, M. L. (2012). *Scheduling: Theory, Algorithms, and Systems* (Vol. 29, 4th ed.). Springer-Verlag. https://doi.org/10.1007/978-1-4614-2361-4

Rippin, D. W. T. (1993). Batch process systems engineering: A retrospective and prospective review. *Computers & Chemical Engineering, 17*, S1–S13.

Saygin, C., & Kilic, S. E. (1999). Integrating flexible process plans with scheduling in flexible manufacturing systems. *The International Journal of Advanced Manufacturing Technology, 15*(4), 268–280.

Srihari, K., & Greene, T. J. (1990). MACRO-CAPP: A prototype CAPP system for an FMS. *The International Journal of Advanced Manufacturing Technology, 5*(1), 34–51.

Sundaram, R. M., & Fu, S.-S. (1988). Process planning and scheduling—A method of integration for productivity improvement. *Computers & Industrial Engineering, 15*(1–4), 296–301.

Tan, W., & Khoshnevis, B. (2000). Integration of process planning and scheduling—A review. *Journal of Intelligent Manufacturing, 11*(1), 51–63.

Wilhelm, W. E., & Shin, H.-M. (1985). Effectiveness of alternate operations in a flexible man-ufacturing system. *International Journal of Production Research*, *23*(1), 65–79.

Yan, H.-S., Xia, Q.-F., Zhu, M.-R., Liu, X.-L., & Guo, Z.-M. (2003). Integrated produc-tion planning and scheduling on automobile assembly lines. *IIE Transactions*, *35*(8), 711–725.

Yang, Y.-N., Parsaei, H. R., & Leep, H. R. (2001). A prototype of a feature-based multiple-alternative process planning system with scheduling verification. *Computers & Industrial Engineering*, *39*(1–2), 109–124.

Yu, M., Zhang, Y., Chen, K., & Zhang, D. (2015). Integration of process planning and sched-uling using a hybrid GA/PSO algorithm. *The International Journal of Advanced Manufacturing Technology*, *78*(1–4), 583–592.

Yuan, B., Zhang, C., Shao, X., & Jiang, Z. (2015). An effective hybrid honey bee mating opti-mization algorithm for balancing mixed-model two-sided assembly lines. *Computers & Operations Research*, *53*, 32–41.

Zhang, H.-C. C., & Mallur, S. (1994). An integrated model of process planning and produc-tion scheduling. *International Journal of Computer Integrated Manufacturing*, *7*(6), 356–364. https://doi.org/10.1080/09511929408944623

Zhao, F., Hong, Y., Yu, D., Yang, Y., & Zhang, Q. (2010). A hybrid particle swarm optimisa-tion algorithm and fuzzy logic for process planning and production scheduling integra-tion in holonic manufacturing systems. *International Journal of Computer Integrated Manufacturing*, *23*(1), 20–39.

6 Dynamic Integrated Process Planning and Scheduling

6.1 INTRODUCTION: BACKGROUND AND DRIVING FORCES

Until this chapter, scheduling problems encountered in the book were considered in a static environment. In addition, most scheduling studies were conducted in a static environment. In static scheduling problems, the system's status, such as the arrival times of jobs and operation times, is known in advance and does not change throughout the scheduling. However, there are dynamic or random events in real job shops, such as new job arrivals, machine breakdowns, and canceling an existing order. Scheduling aims to reduce these uncertainties, adversely affecting the scheduling performance by including random events in the system. This chapter discusses dynamic events in the scheduling problem and dynamic events in the integrated scheduling and process planning problem.

6.2 DYNAMIC ENVIRONMENTS

Scheduling in manufacturing is a problem that has been addressed from various perspectives. There are 14 different types of job shop scheduling problems (JSSP) which are deterministic, dynamic, static, fuzzy, stochastic, no-wait, just-in-time, large-scale, preemptive, flexible, reentrant, assembly, cyclic, and periodic. These are shown in Figure 6.1, with publication numbers between 2015 and 2022. Abdolrazzagh-Nezhad and Abdullah (2017) presented a taxonomy of the JSSP. One of these aspects is related to the system characteristics.

Pinedo (2016) created an essential book on schedule using deterministic and stochastic models. Accordingly, deterministic systems include finite numbers of machines and jobs. In other words, all variables were assumed to be known and not random. Again, there are a finite number of jobs and machines in stochastic systems. Nevertheless, one or more variables must be random in this case, and only their distributions are known in advance. For example, if the process times of the jobs arrive following a distribution, the system behaves stochastically. Other random events included random due dates, job weights, and arrival times.

Xiong et al. (2022) surveyed the types and models of scheduling problems. Based on this, 297 JSSP studies were conducted between 1960 and 2020. From a general perspective, the characteristics of JSSP can be specified according to the state of the system, such as static or dynamic systems. The status of the variables is known at any time in a static system. For example, information such as how many jobs are done/

DOI: 10.1201/9781003215295-6

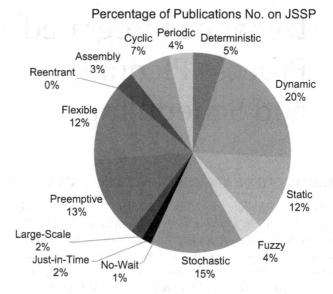

FIGURE 6.1   Number of publications of different JSSP variants.

will be done is known. In dynamic systems, the status of variables at any time is unknown. Over time, changes may occur in the dynamic systems. For example, the number of jobs may change in a dynamic event, in which jobs arrive at any time. Alternatively, the number of machines may change over time in the dynamic event of a machine breakdown. These changes in the properties of the problem also cause changes to the solution.

As in dynamic working environments, clear information regarding the system's state is not known in advance. In addition, dynamic and stochastic production environments have been studied in a more realistic environment if the variables do not behave deterministically. Studies in dynamic and stochastic environments are closest to real production environments. This is because, in real production environments, sometimes, there may be a situation where an unproductive order arrives during the scheduling phase. Alternatively, an urgent job can be expected to arrive and deliver quickly. In this case, the current schedule must be canceled, and decisions such as rescheduling, updating the schedule, and merging existing and new schedules should be made. Some of the dynamic events encountered in a job shop include:

- Arrivals of a job at any time
- Machine breakdown
- Workers' illness
- Changes in delivery times
- Changes in the weight of the work
- Urgent work is coming into the system
- Cancellation of a job
- Other unexpected events

Many job shop scheduling problems (JSSP) are classified in the NP-Hard Many job shop scheduling problems (JSSP) are classified in the NP-Hard problem complexity class. Therefore, even small dynamic and stochastic problems require significant CPU power and computation times. Thus, heuristic algorithms can efficiently solve these problems and achieve good results in fewer trials. Related studies are examined in the next section.

6.3   REVIEW OF THE LITERATURE

6.3.1   RELATED WORKS ON DYNAMIC JSSP

JSSP, an NP-Hard problem, is a field of computational optimization that has been extensively studied, and techniques have been developed for long-term solutions (Motaghedi-Larijani et al., 2010). In recent decades, numerous studies have been conducted on single-machine scheduling, flow shop problems, flexible shop floors, and job shop scheduling. Therefore, it can be said that the JSSP is a saturated area, and it is challenging to find new research areas. Many mathematical deterministic models, mixed-integer models, metaheuristics, multi-agent systems, machine learning, and artificial intelligence techniques have been developed to solve the JSSP, and the best or near-best solutions have been attempted. However, many of these studies can be considered as having a static JSSP environment. The dynamic JSSP problem was first introduced in the literature by Holloway and Nelson (1977). In an early study, Muhlemann et al. (1982) examined a dynamic scheduling problem that works if new jobs arrive in the system. Some dynamic studies conducted before 2000 can be summarized as follows: (Pinedo & Wie, 1986; Rajendran & Holthaus, 1999; Shanthikumar & Buzacott, 1981).

It is believed that static scheduling studies will not yield unintended results in actual job shop environments. s and Liu (1993) presented a critical view of this phenomenon. The study stated that classical scheduling studies were practically poor. They found that deterministic conditions in static studies cannot be applied in real job shops, so that poor-quality schedules will appear. Another study using the same idea was conducted by Cowling and Johansson (2002). The study mentioned different situations between theoretical and practical scheduling. It is impossible to examine real situations using theoretical scheduling. Two principles have been defined to resolve this situation: rescheduling and scheduling of repairs.

When comparing the number of publications on static and dynamic scheduling in the Scopus database, we investigated the number of studies conducted in 2020, 2021, and 2022. Accordingly, the number of studies that mentioned "rescheduling" in the keyword or abstract was 783, "dynamic scheduling" was 697, and "stochastic scheduling" was 144. However, the expression "deterministic scheduling" was provided in 38 studies. This shows that current studies are carried out in rescheduling and dynamic scheduling, as expected. Owing to the increase in studies on dynamic scheduling, literature review studies are needed (Suresh & Chaudhuri, 1993). In the literature, different types of taxonomies exist in stochastic scheduling. Ouelhadj and Petrovic (2009) examined dynamic scheduling in three categories: (i) completely reactive scheduling, in which schedules are applied in real time, and (ii) predictive-reactive scheduling, in which schedules are revised and applied when a dynamic event occurs. (iii) Robust proactive scheduling, wherein schedules are prepared in a dynamic environment to achieve the best performance. Sabuncuoglu and Goren (2009) categorized this taxonomy into reactive and proactive approaches. A more extensive categorization is given as reactive, stochastic, fuzzy, proactive, and sensitivity analyses (Herroelen & Leus, 2005).

At this stage, a literature summary is provided on the dynamic JSSP. Zhang et al. (2013) studied the dynamic JSSP using a hybrid genetic algorithm (GA) and tabu search (TS) to consider machine breakdowns and random job arrivals. Likewise, in another study, considering machine breakdowns and random job arrivals, Zandieh and Adibi (2010) discussed the problem of using dynamic JSSP using an artificial neural network with a neighborhood search algorithm. The results are based on different dispatching rules, such as the shortest processing time (SPT), first-in, first-out (FIFO), and last-in-first-out (LIFO). In a similar study, Kundakci and Kulak (2016) studied a dynamic JSSP to minimize the makespan value using a hybrid genetic algorithm. This study considered new job arrivals, machine breakdowns, and changes in process times. As a result, the developed GA yielded better results than TS.

Cowling and Johansson (2002) investigated single-machine scheduling. This study has 50 frameworks for how real-time information can be used for JSSP. Mason et al. (2004) developed rescheduling strategies to minimize tardiness. In their study, dynamic conditions such as machine breakdowns and the arrival of an urgent job were considered. A complex job shop problem was solved using three different scheduling strategies: right shift, fixed sequence, and total rescheduling. Wong et al. (2006b), using multi-agent systems (MAS), proposed solutions to dynamic JSSP using multi-agent systems. The advantages of the method developed for better global performance are emphasized. The algorithm's efficiency was compared with performance metrics such as overall makespan, mean machines' utilization rate, mean parts' flowtime, and the sum of machines' loading deviation, and MAS structure was observed to work best. Shen and Yao (2015) addressed a multi-objective JSSP with a dynamic, efficient, and stable objective using a multi-objective evolutionary algorithm. In their study, the hypervolume ratio (Van Veldhuizen & Lamont, 1999), generational distance (Van Veldhuizen & Lamont, 1998), distance variance of neighboring vectors, and spread (Deb et al., 2002) metrics were used as performance metrics. The features that should be observed in the literature are the dynamic event, the method used, and the objective function. Information on these studies is presented in Table 6.1.

6.3.2 REVIEW WORKS ON DYNAMIC INTEGRATED PROCESS PLANNING AND SCHEDULING

Dynamic integrated process planning and scheduling (IPPS) is a higher integration of dynamic JSSP. JSSPs are applied separately from process planning, which leads to inefficient performance. To achieve significantly improved overall performance, the IPPS problem mentioned in the previous chapters of the book is of great importance. Therefore, IPPS is a popular and essential field of study. In addition, dynamic events

TABLE 6.1
Summary of the Related Works

Reference	Dynamic Event	Method	Performance Metric
Zhao and Li (2014)	machine breakdown	simulation-based	mean, maximum, variance of tardiness, and proportion of tardy jobs
Yu and Liang (2001)	machine breakdown and new job arrivals	neural networks(NNs) and genetic algorithm(GA)	completion time of all the jobs called the makespan
Liu et al. (2005)	machine breakdown and new job arrivals	metaheuristics - tabu search	makespan
Kunnathur et al. (2004)	new job arrives	simulation-based	shortest processing time (SPT), critical ratio, total work
Han et al. (2021)	machine breakdown	intelligent optimization algorithm	makespan
Goren et al. (2012)	machine breakdowns	branch-and-bound algorithm	stability
Nouiri et al. (2017)	machine breakdowns	particle swarm optimization	makespan, stability
Singh et al. (2007)	machine breakdown	simulation-based	mean flow time, maximum flow time, the variance of flow time, mean tardiness, maximum tardiness, percentage of tardy jobs
Singh et al. (2005)	machine breakdown	simulation-based	mean tardiness, maximum tardiness, and the number of tardy jobs

were added to the IPPS problem to create shop floors that are more suitable for real shop floors. The dynamic events described in the dynamic JSSP section are added to the IPPS problem, and the dynamic IPPS problem is introduced in the literature.

It takes more time to implement and model dynamic and integrated studies in manufacturing systems. Therefore, there has not been much research in this field. Nevertheless, it can be said that integrated and dynamic studies are the most suitable models for real manufacturing systems among the models developed so far. We have included the studies here according to the dates of publication. The first study to be examined was in 2001. The language of the study was Chinese, so only the abstract has been discussed. Their study created optimal process plans and schedules using a genetic algorithm (GA) control mechanism. The problem addressed is "the dynamic integrated process planning and scheduling system" (DIPS). Wong et al. (2006a) focused on rescheduling and process planning. This problem was solved using a multi-agent system (MAS). In this study, an algorithm was proposed to solve the problem of insufficient time in all-scale scheduling. Nejad et al. (2008) used a similar method (MAS) to address dynamic situations in production environments. They first used a heuristic algorithm to create a process plan and schedule and then solved the dynamic IPPS problem using the MAS approach (Nejad et al., 2011). Another MAS operation is performed by Zhang et al. (2012).

Lv and Qiao (2011) have mentioned scheduling difficulties under uncertainty. Dynamic events such as the arrival of new jobs, machine breakdown, and order cancellation events are discussed in this study. Dynamic events increase scheduling complexities and affect scheduling and shop floor efficiencies. Therefore, an evolutionary algorithm has produced solutions that optimize the performance metric for dynamic IPPS problems, and the performance of different algorithms has been compared with case studies.

In Xia et al.'s (2016) study, a hybrid genetic algorithm used variable neighborhood search to study machine breakdowns and new job arrivals as dynamic events to solve the IPPS problem. The system was updated using a simulation study of the developed model according to the dynamic event. The performance metrics of the study are the makespan and mean flow time. In a similar study using the improved GA method, the makespan was improved using MAS with an ant colony optimization approach (Yin et al., 2017).

Jin et al. (2017) developed a mixed-integer linear programming model to optimize the makespan, stability, and tardiness metrics. Experiments reveal that the length of a scheduling interval, the quantity of newly added jobs, and shop utilization significantly affect the efficiency of a production system.

In a recent study, Yu et al. (2018) solved the dynamic IPPS problem using a discrete particle swarm optimization algorithm. In this study, comparative findings are presented to demonstrate the efficiency and effectiveness of the proposed IPPS approach. A scheduling function can specify more than one objective function. Zhang et al. (2020) developed a multi-objective function model. The developed model is solved with a GA and reduced energy usage by approximately 15% for two distinct scheduling schemes with the same makespan, according to the case study results in this study.

6.3.3 Mathematical Models and Performance Metrics

Various mathematical models have been developed to solve dynamic IPPS problems. In some studies, the structure of an intuitive algorithm was mentioned without providing a mathematical model. At this stage, an overview based on two studies on dynamic IPPS is presented. Following the given model, the performance metrics used in the literature are mentioned. Jin et al. (2017) developed a mixed-integer linear programming model to solve dynamic IPPS problems. The assumptions of this study are as follows.

- The machines are instantly ready at t = 0 and can only handle one job.
- There was no job preemption.
- After one operation is completed, the other is immediately switched.
- The setup times were included in the process times.

Second, Xia et al. (2016) make two assumptions. (i) The rescheduling time was assumed to be negligible. (ii) If a dynamic event occurs during an operation, the operation must be reprocessed. The same study discussed dynamic events, such as new job arrivals and machine breakdowns. The following equations are used for new job arrivals:

$$O_{ijkm''} = \{O_{ijk} \,|\, R_{ijkm''} \le T_a\} \rightarrow m = [1,\dots,M]$$

$$O_{ijkm'} = \{O_{ijk} \,|\, R_{ijkm'} \le T_a \le R_{ijkm''}\} \rightarrow m = [1,\dots,M]$$

$$RN_i = R_{ijkm''} \times Z_{ijkm'} + T_a \times (1 - Z_{ijkm'}) \rightarrow i = [1,\dots,N]$$

$$RM_m = R_{ijkm''} \times Z_{ijkm'} + T_a \times (1 - Z_{ijkm'}) \rightarrow m = [1,\dots,M] \tag{6.1}$$

For machine breakdown,

$$RN_i = \begin{cases} R_{ijkm''} \times Z_{ijkm'} + T_{br} \times (1 - Z_{ijkm'}) \rightarrow i = [1,\dots,N], m \ne r \\ (R_{ijkm''} + T_{dr}) \times Z_{ijkm'} + T_{br} \times (1 - Z_{ijkm'}) \rightarrow i = [1,\dots,N], m = r \end{cases} \tag{6.2}$$

$$RN_m = \begin{cases} R_{ijkm''} \times Z_{ijkm'} + T_{br} \times (1 - Z_{ijkm'}) \rightarrow m \ne r \\ (R_{ijkm''} + T_{dr}) \times Z_{ijkm'} + T_{br} \times (1 - Z_{ijkm'}) \rightarrow m = r \end{cases} \tag{6.3}$$

$$Z_{ijkm'} = \begin{cases} 1 \text{ if the machine of operation } O_{ijkm'} \text{ did not change} \\ 0, \text{ otherwise} \end{cases} \tag{6.4}$$

where, $O_{ijkm''}$ the set of completed operations which is performed on machine m when rescheduling, O_{ijk} the kth operation in the jth alternative process plan of job i, $R_{ijkm''}$ the completion time of O_{ijk} in the initial scheduling, T_a the arrival time of the new job, $O_{ijkm'}$ the set of operations that are being machined on machine m when rescheduling, $R_{ijkm'}$ the starting time of O_{ijk} in the initial scheduling, O_{ijk} the kth operation in the jth alternative process plan of job i, $R_{ijkm'}$ the starting time of O_{ijk} in the initial scheduling, RN_i the release time of job i when rescheduling, RM_m the release time of machine m when rescheduling.

An essential component of scheduling problems is the performance metric. Different performance criteria were used in most of the studies above. The performance metrics must be in accordance with the structure of the problem. This section provides brief information about common performance metrics, such as makespan, tardiness, and stability in dynamic IPPS problems.

Singh et al. (2007) observed model performance with two objective functions. The first function is flow-time-based performances where a number of jobs (N), flow time of the job i (F_i), mean flow time (F_{mean}), maximum flow time (F_{max}) and variance of flow time (σ_F^2) calculated as follows:

$$F_i = \frac{1}{N}\left(\sum_{i=1}^{i=N} F_i\right) \quad F_{max} = \max(F_i, 1 < i < N) \quad \sigma_F^2 = \frac{\sum_{i=1}^{i=N}(F_i - F_{mean})^2}{N} \tag{6.5}$$

The second function is tardiness-based performance metrics, where tardiness of the job i (T_i), mean tardiness ($\bar{T}$), maximum tardiness (T_{max}), percentage of tardy jobs ($\%N_t$) calculated as follows:

$$\bar{T} = \frac{\sum_{i=1}^{i=N} T_i}{N} T_{\max} = \max(T_i, 1 < i < N)\%N_t = \frac{i, T_i > 0, 1 < i < N}{N} \times 100 \qquad (6.6)$$

The most used measure of performance in scheduling work is makespan (C_{max}). This is calculated as follows:

$$C_{\max} = \max\left(t_{ij}\right) \qquad (6.7)$$

where t_{ij} is the completion time of operations.

Rangsaritratsamee et al. (2004) used a stability performance metric with two components: (i) the starting time total deviation (for all such jobs between the starting times in a new schedule and the old schedule); (ii) penalty with rescheduling jobs as follows (Stability = Starting time deviation + total deviation penalty) (Fattahi & Fallahi, 2010):

$$\text{stability} = \sum_{\text{all eligible } n} \sum_{m} \left| t'_{n,m} - t_{n,m} \right| + \sum_{\text{all eligible } n} \sum_{m} PF \left| t'_{n,m} + t_{n,m} - 2t \right| \qquad (6.8)$$

Where, $t_{n,m}$ is starting time of job n on the machine m per a schedule built at one rescheduling point, $t'_{n,m}$ Schedule built at the next rescheduling point t, current time, PF is a penalty function associated with total deviation from the current time.

6.4   CONCLUSION

This chapter discusses advanced topics of scheduling problems, dynamic DSSP, and dynamic IPPS. First, a literature review is provided, and then mathematical models and performance metrics are mentioned. Metaheuristic algorithms and solution methods for the problems above are provided in Chapter 11. Therefore, the solution algorithms are not included in this chapter.

REFERENCES

Abdolrazzagh-Nezhad, M., & Abdullah, S. (2017). Job shop scheduling: Classification, constraints and objective functions. *International Journal of Computer, Electrical, Automation, Control and Information Engineering, 11*(4), 6.

Cowling, P., & Johansson, M. (2002). Using real time information for effective dynamic scheduling. *European Journal of Operational Research, 139*(2), 230–244. https://doi.org/10.1016/S0377-2217(01)00355-1

Deb, K., Pratap, A., Agarwal, S., & Meyarivan, T. (2002). A fast and elitist multi-objective genetic algorithm: NSGA-II. *IEEE Transactions on Evolutionary Computation, 6*(2), 182–197.

Fattahi, P., & Fallahi, A. (2010). Dynamic scheduling in flexible job shop systems by considering simultaneously efficiency and stability. *CIRP Journal of Manufacturing Science and Technology, 2*(2), 114–123. https://doi.org/10.1016/j.cirpj.2009.10.001

Goren, S., Sabuncuoglu, I., & Koc, U. (2012). Optimization of schedule stability and efficiency under processing time variability and random machine breakdowns in a job shop environment. *Naval Research Logistics (NRL)*, *59*(1), 26–38. https://doi.org/10.1002/nav.20488

Han, D., Li, W., Li, X., Gao, L., & Li, Y. (2021). A data-driven proactive scheduling approach for hybrid flow shop scheduling problem. *International Manufacturing Science and Engineering Conference*, *85079*, V002T07A002.

Herroelen, W., & Leus, R. (2005). Project scheduling under uncertainty: Survey and research potentials. *European Journal of Operational Research*, *165*(2), 289–306.

Jin, L., Zhang, C., Shao, X., & Yang, X. (2017). A study on the impact of periodic and event-driven rescheduling on a manufacturing system: An integrated process planning and scheduling case. *Proceedings of the Institution of Mechanical Engineers, Part B: Journal of Engineering Manufacture*, *231*(3), 490–504. https://doi.org/10.1177/095440541662 9585

Kundakcı, N., & Kulak, O. (2016). Hybrid genetic algorithms for minimizing makespan in dynamic job shop scheduling problem. *Computers & Industrial Engineering*, *96*, 31–51. https://doi.org/10.1016/j.cie.2016.03.011

Kunnathur, A. S., Sundararaghavan, P. S., & Sampath, S. (2004). Dynamic rescheduling using a simulation-based expert system. *Journal of Manufacturing Technology Management*, *15*(2), 199–212.

Liu, S. Q., Ong, H. L., & Ng, K. M. (2005). Metaheuristics for minimizing the makespan of the dynamic shop scheduling problem. *Advances in Engineering Software*, *36*(3), 199–205. https://doi.org/10.1016/j.advengsoft.2004.10.002

MacCarthy, B. L., & Liu, J. (1993). Addressing the gap in scheduling research: A review of optimization and heuristic methods in production scheduling. *The International Journal of Production Research*, *31*(1), 59–79.

Mason, S. J., Jin, S., & Wessels, C. M. (2004). Rescheduling strategies for minimizing total weighted tardiness in complex job shops. *International Journal of Production Research*, *42*(3), 613–628. https://doi.org/10.1081/00207540310001614132

Motaghedi-Larijani, A., Sabri-Laghaie, K., & Heydari, M. (2010). Solving flexible job shop scheduling with multi objective approach. *International Journal of Industrial Engineering & Production Research*, *21*(4), 197–209.

Muhlemann, A. P., Lockett, A. G., & Farn, C.-K. (1982). Job shop scheduling heuristics and frequency of scheduling. *The International Journal of Production Research*, *20*(2), 227–241.

Nejad, H. T. N., Sugimura, N., & Iwamura, K. (2011). Agent-based dynamic integrated process planning and scheduling in flexible manufacturing systems. *International Journal of Production Research*, *49*(5), 1373–1389. https://doi.org/10.1080/00207543.2010.518741

Nejad, H. T. N., Sugimura, N., Iwamura, K., & Tanimizu, Y. (2008). Integrated dynamic process planning and scheduling in flexible manufacturing systems via autonomous agents. *Journal of Advanced Mechanical Design, Systems, and Manufacturing*, *2*(4), 719–734. https://doi.org/10.1299/jamdsm.2.719

Nelson, R. T., Holloway, C. A., & Mei-Lun Wong, R. (1977). Centralized scheduling and priority implementation heuristics for a dynamic job shop model. *AIIE Transactions*, *9*(1), 95–102.

Nouiri, M., Bekrar, A., Jemai, A., Trentesaux, D., Ammari, A. C., & Niar, S. (2017). Two stage particle swarm optimization to solve the flexible job shop predictive scheduling problem considering possible machine breakdowns. *Computers & Industrial Engineering*, *112*, 595–606. https://doi.org/10.1016/j.cie.2017.03.006

Ouelhadj, D., & Petrovic, S. (2009). A survey of dynamic scheduling in manufacturing systems. *Journal of Scheduling*, *12*(4), 417–431. https://doi.org/10.1007/s10951-008-0090-8

Pinedo, M. L. (2016). *Scheduling*. Springer International Publishing. https://doi.org/10.1007/978-3-319-26580-3

Pinedo, M., & Wie, S.-H. (1986). Inequalities for stochastic flow shops and job shops. *Applied Stochastic Models and Data Analysis, 2*(1–2), 61–69.

Qiao, L. H., & Lv, S. P. (2011). Models and implementation of integrated process planning and scheduling. *Advanced Materials Research, 314–316,* 518–523. https://doi.org/10.4028/www.scientific.net/AMR.314-316.518

Rajendran, C., & Holthaus, O. (1999). A comparative study of dispatching rules in dynamic flowshops and jobshops. *European Journal of Operational Research, 116*(1), 156–170.

Rangsaritratsamee, R., Ferrell, W. G., & Kurz, M. B. (2004). Dynamic rescheduling that simultaneously considers efficiency and stability. *Computers & Industrial Engineering, 46*(1), 1–15. https://doi.org/10.1016/j.cie.2003.09.007

Sabuncuoglu, I., & Goren, S. (2009). Hedging production schedules against uncertainty in manufacturing environment with a review of robustness and stability research. *International Journal of Computer Integrated Manufacturing, 22*(2), 138–157.

Shanthikumar, J. G., & Buzacott, J. A. (1981). Open queueing network models of dynamic job shops. *The International Journal of Production Research, 19*(3), 255–266.

Shen, X.-N., & Yao, X. (2015). Mathematical modeling and multi-objective evolutionary algorithms applied to dynamic flexible job shop scheduling problems. *Information Sciences, 298,* 198–224. https://doi.org/10.1016/j.ins.2014.11.036

Singh, A., Mehta, N. K., & Jain, P. K. (2005). Tardiness based new dispatching rules for shop scheduling with unreliable machines. *International Journal of Simulation Modelling, 4*(1), 5–16.

Singh, A., Mehta, N. K., & Jain, P. K. (2007). Multicriteria dynamic scheduling by swapping of dispatching rules. *The International Journal of Advanced Manufacturing Technology, 34*(9–10), 988–1007. https://doi.org/10.1007/s00170-006-0674-4

Suresh, V., & Chaudhuri, D. (1993). Dynamic scheduling—A survey of research. *International Journal of Production Economics, 32*(1), 53–63. https://doi.org/10.1016/0925-5273(93)90007-8

Van Veldhuizen, D. A., & Lamont, G. B. (1998). *Multi-Objective Evolutionary Algorithm Research: A History and Analysis.* Citeseer.

Van Veldhuizen, D. A., & Lamont, G. B. (1999). Multi-objective evolutionary algorithm test suites. *Proceedings of the 1999 ACM Symposium on Applied Computing,* 351–357.

Wong, T. N., Leung, C. W., Mak, K. L., & Fung, R. Y. K. (2006a). Integrated process planning and scheduling/rescheduling—An agent-based approach. *International Journal of Production Research, 44*(18–19), 3627–3655. https://doi.org/10.1080/00207540600675801

Wong, T. N., Leung, C. W., Mak, K. L., & Fung, R. Y. K. (2006b). Dynamic shopfloor scheduling in multi-agent manufacturing systems. *Expert Systems with Applications, 31*(3), 486–494. https://doi.org/10.1016/j.eswa.2005.09.073

Xia, H., Li, X., & Gao, L. (2016). A hybrid genetic algorithm with variable neighborhood search for dynamic integrated process planning and scheduling. *Computers & Industrial Engineering, 102,* 99–112. https://doi.org/10.1016/j.cie.2016.10.015

Xiong, H., Shi, S., Ren, D., & Hu, J. (2022). A survey of job shop scheduling problem: The types and models. *Computers & Operations Research, 142,* 105731. https://doi.org/10.1016/j.cor.2022.105731

Yin, L., Gao, L., Li, X., & Xia, H. (2017). An improved genetic algorithm with rolling window technology for dynamic integrated process planning and scheduling problem. *2017 IEEE 21st International Conference on Computer Supported Cooperative Work in Design (CSCWD),* 414–419. https://doi.org/10.1109/CSCWD.2017.8066730

Yu, H., & Liang, W. (2001). Neural network and genetic algorithm-based hybrid approach to expanded job-shop scheduling. *Computers & Industrial Engineering, 39*(3–4), 337–356.

Yu, M. R., Yang, B., & Chen, Y. (2018). Dynamic integration of process planning and scheduling using a discrete particle swarm optimization algorithm. *Advances in Production Engineering & Management, 13*(3), 279–296. https://doi.org/10.14743/apem2018.3.290

Zandieh, M., & Adibi, M. A. (2010). Dynamic job shop scheduling using variable neighbour-hood search. *International Journal of Production Research*, *48*(8), 2449–2458. https://doi.org/10.1080/00207540802662896

Zhang, L., Gao, L., & Li, X. (2013). A hybrid genetic algorithm and tabu search for a multi-objective dynamic job shop scheduling problem. *International Journal of Production Research*, *51*(12), 3516–3531. https://doi.org/10.1080/00207543.2012.751509

Zhang, L., Wong, T. N., & Fung, R. Y. K. (2012). A multi-agent system for dynamic integrated process planning and scheduling using heuristics. In G. Jezic, M. Kusek, N.-T. Nguyen, R. J. Howlett, & L. C. Jain (Eds.), *Agent and Multi-Agent Systems. Technologies and Applications* (pp. 309–318). Springer. https://doi.org/10.1007/978-3-642-30947-2_35

Zhang, X., Zhang, H., & Yao, J. (2020). Multi-objective optimization of integrated process planning and scheduling considering energy savings. *Energies*, *13*(23), 6181. https://doi.org/10.3390/en13236181

Zhao, F. Q., & Li, N. (2014). Flow time and tardiness based on new scheduling rules for dynamic shop scheduling with machine breakdown. *Applied Mechanics and Materials*, *556*, 4412–4416.

7 Scheduling with Due-Date Assignment

7.1 INTRODUCTION

The scheduling problem with the due-date assignment has been a popular study topic for the last decades. In the past, the due dates were dictated from the outside, unaware of the shop floor status. Later, the firm was trying to comply with these unrealistic due dates. Of course, the due dates given without having information about the shop floor did not adhere to the shop floor realities, and as a result, the unreasonable due dates were often not complied with. Such a due-date setting is called an exogenous due-date assignment.

Also, classically only tardiness was penalized, excluding the costs of earliness, storage, spoilage, and the cost of capital allocation into stock. With the concepts of Just-in-Time and Lean Manufacturing that emerged and settled later, earliness has also become undesirable. Although earliness was not desired, tardiness remained more undesirable than earliness Figure 7.1 and Table 7.1.

Thus, instead of externally dictated due dates, Firms began to use jointly determined times with the customer and even determined within the firm according to the status of the shop floor. Since the due dates are given inside, taking into account the status of the shop floor, there can now be more realistic and more advantageous due dates can be given for the firm.

As a result of assigning more reasonable due dates, it became possible to lower the cost of early completion and tardiness costs. Thus, the overall penalty function became even lower. Since only job shop scheduling is an NP-Hard problem, the scheduling problem with due-date assignment is also in the class of NP-Hard problems for the job shop environment.

7.2 REVIEW OF THE LITERATURE

Many studies have been conducted in the last decades on scheduling and due-date determination. Scheduling with due-date determination is a prevalent research topic in the literature (Gordon et al., 2002, 2012). The most commonly studied problem on this subject, which has been studied for decades, is single-machine problems (Wang et al., 2016; Zhao et al., 2018; Kacem and Kellerer, 2019).

Although single-machine scheduling and due-date assignment problems are studied intensively, there are also studies conducted for environments other than single-machine environments. It is possible to encounter studies in the literature for two machine environments, as in (Birman and Mosheiov, 2004; Sakuraba et al., 2009). There are also studies on the flow shop environment (Geng et al., 2019; Koulamas and Kyparisis,

DOI: 10.1201/9781003215295-7

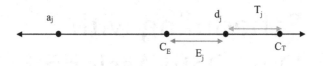

FIGURE 7.1   Scheduling with due-date assignment.

TABLE 7.1

Notations

p_j, a_j = Total Processing Time / Arrival time of job j	C_j Completion time (C_E = Early completion, C_T = Tardy Completion)
d_j = Due-date of job j	$F_j = C_j - a_j$ Flow Time of job j
$\forall\, d_j = d$ = Common due-date	$E_j = \max\{0, d_j - C_j\}$ Earliness of Job j
$W_j = d_j - a_j - p_j$ Waiting time of job j	$T_j = \max\{0, C_j - d_j\}$ Tardiness of Job j
$l_j = d_j - a_j$ Lead Time of job j	$L_j = C_j - d_j$ Lateness

---

2022). The two machine environments mentioned here are two machine flow shop environments.

It is also possible to find many studies on the parallel machine environment (Kim et al., 2012; Yin et al., 2014). Also, for job shop-type production following examples can be found (Lauff and Werner, 2004; Chiang and Fu, 2007; Baykasoğlu and Göçken, 2009; Vinod and Sridharan, 2011).

In addition to the literature briefly mentioned above, the following studies can be given as examples, focusing on current literature. These studies are presented in more detail. Gordon et al. (2012) addressed the problem of scheduling and due-date determination for special cases in process times and did a brief literature review in this area. They discussed cases where precedence relationships exist, process times are controllable, learning and deterioration effects possible, and maintenance activity affects machine performance.

As mentioned before, scheduling with due-date assignment (SWDDA) problems has been done for many machine environments, and it is said that most of them were done for single-machine environments. These studies are discussed in more detail below.

Y. Yin et al. (2013) studied the problem of single-machine scheduling and common due-date determination problem where the process times are a convex function that depends on the amount of continuously divisible resources assigned to those jobs. Xingong and Yong (2015) studied single-machine scheduling with common (CON) and slack (SLK) due-window assignment problem. Here, the due window assignment approach is used instead of the due date assignment, and this approach will be discussed in detail in Chapter 8. In this study, the processing times of the jobs depend on the total processing time of the jobs processed so far, and there is a learning effect.

D.-J. Wang et al. (2016) studied single-machine scheduling and due-date assignment problem in which two agents try to optimize their performance. Three due-date assignment models were considered. These are common, slack, and unrestricted due-date assignment methods. Zhao et al. (2018) studied the single-machine scheduling problem with due date determination. In their studies, CON, where a common due date is given for each job, and SLK, where different due dates are provided for each job, were used. In this study, the process times of the jobs are determined depending on the beginning and the sequence of the jobs. In this study, as a penalty function, the sum of the cost of changing the due date, the cost of the discarded jobs and the early completion costs were tried to be minimized.

Xiong et al. (2018) studied the problem of single-machine scheduling with common due-date assignment in case of potential machine disruption. Kacem and Kellerer (2019) studied the problem of single-machine scheduling with a common due-date assignment for situations where the process times are uncertain and within a range depending on the job.

Birman and Mosheiov (2004) discussed the due-date assignment and scheduling problem for two machine flow shop environments. S. Sakuraba et al. (2009) studied the job scheduling problem in two machine flow shop environments and tried to minimize the mean of absolute deviation from the common due date as an objective function. Mosheiov and Sarig (2009) studied the scheduling and due-date assignment problem in a parallel uniform machines environment for a two-machine case.

Geng et al. (2019) discussed the common due-date assignment and scheduling problem in the no-wait flow shop environment. In this problem, it is assumed that there is a convex resource allocation and learning effect in two machine environments. It is assumed that the processing time of each job depends on its position and the extra resource allocated. Koulamas and Kyparisis (2022) studied the flow shop scheduling problem for two different job due dates. Here, the problem where the jobs belong to two different customers and a separate due date is assigned for each customer is discussed.

Kim et al. (2012) studied the scheduling and common due-date assignment problem in a parallel machine environment. Three decision variables were accepted, and they tried to decide on allocating jobs to parallel machines, the sequencing of jobs in front of each machine, and a common due date. As the objective function, the sum of the due-date assignment, earliness, and tardiness penalties has been tried to be minimized.

N. Yin et al. (2014) studied the unrelated parallel machine scheduling problem. In this study, it is assumed that the process times depend on resource allocation. There are deteriorating jobs, so the process times are determined based on the job's start time and resource allocation. As the objective function, it was tried to decide on the optimum work order and resource allocation.

Lauff and Werner (2004) discussed scheduling with a common due date in the shop and multi-machine environment. They discussed where common due dates, earliness, and tardiness were penalized and conducted a survey on this field. Chiang and Fu (2007) studied dispatching rules in a job shop environment. Due-date-based objectives such as tardy rate, average tardiness, and maximum tardiness were used as objective functions, and eighteen rules selected from the literature were considered as dispatching rules.

Vinod and Sridharan (2011) analyzed the due-date assignment methods and scheduling rules together in the dynamic job shop production system. The jobs are assumed to come to the shop floor according to the exponential distribution. They used Dynamic processing plus waiting time (DPPW), Total work content (TWK), Dynamic total work content (DTWK), and Random work content (RWK) methods as due-date determination rules. They used seven different rules as scheduling rules. They found dynamic due-date assignment methods to be successful among the due-date assignment methods.

7.3 MATHEMATICAL SOLUTIONS

Mathematical models and metaheuristics are used to solve this problem. If we look at the current studies in the literature on this subject, the following studies can be given as examples of mathematical solutions. Before giving the literature, it would be useful to give the notations commonly used (Table 7.2).

Kim et al. (2012) studied the problem of assigning a common due date in a parallel machine environment. In this study, three decision variables are defined: common due date to be assigned, allocation of jobs to parallel machines, and sequencing of jobs allocated to each machine. The mathematical model developed for the solution to this problem is given in the study.

TABLE 7.2
Common Notations Used in the Literature

P_{ji}	a convex function of processing time concerning u_{ji}	V_j = Earliness indicator	(if $C_j < d_j$ $V_j = 1$, otherwise 0)
α, α_j	The earliness cost for job J_j / Due-date penalty coefficient of job J_j	U_j = Tardiness indicator	(if $C_j > d_j$ $U_j = 1$, otherwise 0)
γ, γ_j	The Tardiness cost for job J_j / Due-date time length cost	E_j = Earliness	E_j = Max $\{d_j - C_j, 0\}$
$\bar{p}_j, \bar{P}_{ji}$	Processing time of job J_j without resource allocating	T_j Tardiness	T_j = Max $\{C_j - d_j, 0\}$
β, β_j	The Earliness /Tardiness cost for job J_j	c_j	Unit due-date length cost
$d = (d_1,...,d_n)$	Due-dates of the jobs	r^a	a = learning effect
v_j, G_j, g_{ji}	Unit RCC for job J_j	$J = (J_1,..,J_n)$	Set of jobs
u, u_j, u_{ji}	Resource allocated to job J_j	n ($n \geq 2$)	Number of jobs
$Q, R, \hat{G}$	The upper limit on RCC	S, π	Job sequence
K	The upper limit on scheduling cost	$[j]$	Jth position,
w_j	Tardiness cost of job j if there is tardiness	r	rth position
A_j	Acceptable lead time of jth job	m	Positive constant

Shen and Yao (2015) applied mathematical modeling and evolutionary algorithms to the dynamic-flexible job shop scheduling problem. They tried to minimize the function min $F = [f_1(t_1), f_2(t_1), f_3(t_1), f_4(t_1)]$. Here $t_1 > t_0$ is the rescheduling time. $f_1(t_1)$ means minimizing makespan, $f_2(t_1)$ means minimizing tardiness penalty, $f_3(t_1)$ means minimizing maximum machine load, and $f_4(t_1)$ means minimizing the difference between the original and the new schedule.

Instead of the predetermined due dates, Shabtay (2016) studied the problem of assigning due dates that will be different for each job and integrated with the scheduling. In this study, he focused on a situation where each due date would not exceed the announced dates, and these due dates would be determined and integrated with scheduling. In this study, weighted earliness, tardiness, and due-date assignment costs were tried to be minimized. The logic in penalizing the due dates is that the firm can reduce prices because of long due dates. Their study aims to find the schedule (S) and due dates $d = (d_1, d_2, ..., d_n)$ that minimize the total weighted earliness, tardiness, and due-date assignment costs. The objective function is as follows;

$$\text{Min } Z(S, d) = \sum_{j=1}^{n} \left(\alpha_j \max\{0, d_j - A_j\} + \beta_j E_j + \gamma_j T_j \right).$$

Xiong et al. (2018) studied single-machine scheduling with common due date assignment problems in a disruptive environment. In this problem, it is known when the machines will not be used in advance, but when the time comes, it is not known how much they will not be used, which corresponds to a probability distribution. In this study, both the non-resumable and resumable cases are discussed. The dynamic programming method was used in problem-solving. As a penalty function, the sum of early completion, tardiness, and due-date assignment costs has been tried to be minimized. The most appropriate schedule and common due date are tried to be determined to minimize the penalty function.

J.-B. Wang et al. (2018) have extended the problems related to the classical CON and SLK due-date assignment methods with models that include resource allocation and job-related learning effects. They developed several algorithms in their studies to find the optimal order, resource allocation, processing times, and due dates to minimize the objective function. Simple polynomial-time solutions are introduced for all versions examined in the study.

Karhi and Shabtay (2018) have solved single-machine scheduling problems for cases where both due dates and process times are treated as variables. Process times are considered a convex function that decreases depending on the amount of non-renewable resource allocated to this job. Their study discusses CON and SLK due date determination methods, which are very common in the literature. The problem of $1|X, \text{conv}|(Z, V)$ for $X \in \{\text{CON, SLK}\}$ has been tried to be solved. In this study, three versions of the problem have been handled, and the $F(Z, V) = \sum_{j=1}^{n} w_j U_j + \sum_{j=1}^{n} c_j d_j + \sum_{j=1}^{n} v_j u_j$ problem has been tried to be minimized in the first version. In the first version, weighted tardiness, due date determination, and resource usage costs were tried to be minimized. This problem can be called the $1|X, \text{conv}|F(Z, V)$ problem. In the second version, the min $Z(S) = \sum_{j=1}^{n} w_j U_j + \sum_{j=1}^{n} c_j d_j$ subject to $V(S) = \sum_{j=1}^{n} v_j u_j <= R$ problem is addressed. This problem can be represented as $1|X, \text{conv}| \epsilon (Z/V)$. Here, the weighted tardiness and due date costs are tried to be minimized, provided that the resource allocation cost is lower than R. In the third version, the min $V(S) = \sum_{j=1}^{n} v_j u_j$ subject to $Z(S) = \sum_{j=1}^{n} w_j U_j + \sum_{j=1}^{n} c_j d_j <= K$ problem is addressed. This problem can be represented as $1|X, \text{conv}| \in (V/Z)$. In this problem, the resource allocation cost has been tried to be minimized, provided that the weighted tardiness and due date cost are less than K.

Drwal (2018) studied the single-machine scheduling problem, in which the sum of the weighted tardy jobs is minimized. It is assumed here that the due dates are not known precisely, and the decision maker needs to create a schedule. Although the due dates are not known, it is assumed that the interval is known. The concept of maximum regret is used to obtain robust solutions. For the assumption that the weights of the jobs are the same, a polynomial time algorithm is given, and for the general case, a mixed integer programming formulation is presented.

Geng et al. (2019) studied the common due date scheduling problem, in which the learning effect and convex resource allocation in a two-machine no-wait flow shop determine the process times of the jobs. They tried to find the optimum common due date, optimum resource allocation, and optimum schedule that minimizes the total early completion, tardiness, and due date costs (resource allocation cost) when there is a limit to the total resource allocation cost. They developed polynomial time algorithms for two versions of the problem. Here there are n non-preemption independent jobs $\hat{J} = \{J_1, J_2, \dots, J_n\}$ and these jobs will be processed in a non-preemption two-machine flow shop. Each job will be processed first on machine M_1 and then on machine M_2. Operation O_{ji} ($j = 1, 2, \dots, n; i = 1, 2$) of each job will be processed on two consecutive machines without interruption. Each job is ready at time zero.

$$P_{ji} = \left(\frac{\bar{P}_{ji} r^a}{u_{ji}} \right)^m$$

Here m is a positive constant. $\bar{P}_{ji}$ is the normal processing time of O_{ji} operation. u_{ji} is the non-renewable, continuously divisible resource allocated to the operation O_{ji}. P_{ji} is a convex function concerning u_{ji}. Here, the aim is to determine the optimum resource allocation strategy u, the optimum common due-date d, and the optimum schedule π. In this problem, the following function is minimized.

$$\text{Minimize } f(\pi, u, d) = \sum_{j=1}^{n} (\alpha E_j + \beta T_j + \gamma d)$$

Subject to

$$\sum_{i=1}^{2} \sum_{j=1}^{n} g_{ji} u_{ji} \leq \hat{G}$$

where the weights are $\alpha \geq 0, \beta \geq 0, \gamma \geq 0$;

g_{ji} is the resource allocation cost for a unit of time.

7.4 METAHEURISTIC SOLUTIONS

Scheduling jobs with different earliness and tardiness penalties on a single machine by a common due date is an NP-Hard optimization problem, and mathematical solutions are only suitable for small problems.

Min and Cheng (2006) studied the optimum scheduling and common due-date assignment problems in a parallel machine environment. They tried to minimize the total earliness, tardiness, and due-date assignment costs as an objective function. They used Genetic Algorithm (GA), Simulated Annealing-based Genetic Algorithm (GASA), Iterative heuristic fine-tuning operator-based genetic algorithm (GAH), and Simulated Annealing and Iterative heuristic fine-tuning operator-based genetic algorithm (GASAH) metaheuristics as solution methods.

Lin et al. (2007) studied the single machine scheduling problem of the jobs with a common due date and tried to minimize the total earliness and tardiness costs as an objective function. Genetic algorithms (GA) and simulated annealing (SA) metaheuristics using the greedy local search method were used as a solution method.

Nearchou (2008) solved the single-machine scheduling problem with the differential evolution (DE) algorithm according to different early completion and tardiness penalties.

Baykasoğlu and Göçken (2009) proposed gene expression programming (GEP) technique based on genetic programming. GEP is a recently developed member of the genetic programming family. They compared the proposed due-date assignment model with previously used conventional methods.

Zhang et al. (2013) studied the job shop scheduling problem that minimizes the overall weighted tardiness for predetermined due dates. They used the hybrid artificial bee colony algorithm as the solution method. In studies conducted in this area, the makespan $C_{\max} = \max_{j=1}^{n} \{C_j\}$, maximum lateness $L_{\max} = \max_{j=1}^{n} \{L_j\}$, Total tardiness (TT = $\sum_{j=1}^{n} T_j$), Total weighted tardiness (TWT = $\sum_{j=1}^{n} w_j T_j$) are used as a performance measure. Here, C_j is the completion time of the job j, $L_j = C_j - d_j$ is the lateness, and the tardiness is expressed as $T_j = \max \{0, C_j - d_j\}$. While the makespan criterion was used many times in the past, as make-to-order (MTO) has become common, performance measures related to due dates have become important. In this environment, customers are given due-dates and then jobs are scheduled. However, the problem in which the due-date and scheduling are handled in an integrated manner is the problem discussed in this chapter. Zhang et al. (2013) studied the job shop-type scheduling problem where the due dates are decided in advance.

Erden et al. (2019) studied dynamically integrated process planning, scheduling, and due-date assignment. In this study, Genetic Algorithms (GA), Tabu Search (TS), Simulated Annealing (SA), Genetic-Tabu Hybrid Algorithm (GA/TS), Genetic-Simulated Annealing Hybrid algorithms (GA/SA) were used. In this study, jobs come to the shop floor dynamically in accordance with the exponential distribution, and process plan selection is integrated in addition to scheduling and due-date integration.

7.5 CONCLUSION

Scheduling and due-date assignment are two essential production planning functions. They have significant interactions with each other. If these two functions are handled independently, very inefficient solutions may emerge.

Traditionally, these functions were handled separately and independently given due dates, and scheduling could yield poor solutions. Due dates are given internally and externally; classically, these dates were often given without the shop floor's knowledge and did not comply with the shop floor's conditions. Recently, the integration of these two functions has been investigated in many studies, and due dates have been given by considering the shop floor conditions. In this way, the dates given have significantly contributed to the global solution, and better solutions have begun to be obtained.

Again, while classically tardiness was punished, with the establishment of JIT philosophy, earliness also started to be undesirable. In addition, many studies have tried to minimize the cost of assigning a due date along with earliness and tardiness. Demir and Erden (2020) also penalized the length of the due date. Since long due dates will not be preferred by any customer, the length of the due date is penalized in the same way. In addition, in many studies, other objective functions such as total tardiness, maximum tardiness, mean tardiness, and makespan were used as objective functions.

Most of the studies performed are about single-machine problems. Studies are also available for two-machine environments, flow shops, parallel machines, multi-machines, and job shop machining environments.

REFERENCES

Baykasoğlu, A., & Göçken, M. (2009). Gene expression programming based due date assignment in a simulated job shop. *Expert Systems with Applications*, *36*(10), 12143–12150.

Birman, M., & Mosheiov, G. (2004). A note on a due-date assignment on a two-machine flow-shop. *Computers & Operations Research*, *31*(3), 473–480. https://doi.org/10.1016/S0305-0548(02)00225-3

Chiang, T. C., & Fu, L. C. (2007). Using dispatching rules for job shop scheduling with due date-based objectives. *International Journal of Production Research*, *45*(14), 3245–3262. https://doi.org/10.1080/00207540600786715

Demir, H. I., & Erden, C. (2020). Dynamic integrated process planning, scheduling and due-date assignment using ant colony optimization. *Computers & Industrial Engineering*, *149*, 106799. https://doi.org/10.1016/j.cie.2020.106799

Drwal, M. (2018). Robust scheduling to minimize the weighted number of late jobs with interval due-date uncertainty. *Computers & Operations Research*, *91*, 13–20. https://doi.org/10.1016/j.cor.2017.10.010

Erden, C., Demir, H. I., & Kökçam, A. H. (2019). Solving integrated process planning, dynamic scheduling, and due date assignment using metaheuristic algorithms. *Mathematical Problems in Engineering*, *2019*, 1–19. https://doi.org/10.1155/2019/1572614

Geng, X.-N., Wang, J.-B., & Bai, D. (2019). Common due date assignment scheduling for a no-wait flowshop with convex resource allocation and learning effect. *Engineering Optimization*, *51*(8), 1301–1323. https://doi.org/10.1080/0305215X.2018.1521397

Gordon, V., Proth, J.-M., & Chu, C. (2002). A survey of the state-of-the-art of common due date assignment and scheduling research. *European Journal of Operational Research*, *139*(1), 1–25. https://doi.org/10.1016/S0377-2217(01)00181-3

Gordon, V., Strusevich, V., & Dolgui, A. (2012). Scheduling with due date assignment under special conditions on job processing. *Journal of Scheduling*, *15*(4), 447–456. https://doi. org/10.1007/s10951-011-0240-2

Kacem, I., & Kellerer, H. (2019). Complexity results for common due date scheduling problems with interval data and minmax regret criterion. *Discrete Applied Mathematics*, *264*, 76–89. https://doi.org/10.1016/j.dam.2018.09.026

Karhi, S., & Shabtay, D. (2018). Single machine scheduling to minimise resource consumption cost with a bound on scheduling plus due date assignment penalties. *International Journal of Production Research*, *56*(9), 3080–3096. https://doi.org/10.1080/00207543. 2017.1400708

Kim, J.-G., Kim, J.-S., & Lee, D.-H. (2012). Fast and meta-heuristics for common due-date assignment and scheduling on parallel machines. *International Journal of Production Research*, *50*(20), 6040–6057. https://doi.org/10.1080/00207543.2011.644591

Koulamas, C., & Kyparisis, G. J. (2022). Flow shop scheduling with two distinct job due dates. *Computers & Industrial Engineering*, *163*, 107835. https://doi.org/10.1016/j.cie. 2021.107835

Lauff, V., & Werner, F. (2004). Scheduling with common due date, earliness and tardiness penalties for multimachine problems: A survey. *Mathematical and Computer Modelling*, *40*(5–6), 637–655. https://doi.org/10.1016/j.mcm.2003.05.019

Lin, S.-W., Chou, S.-Y., & Chen, S.-C. (2007). Meta-heuristic approaches for minimizing total earliness and tardiness penalties of single-machine scheduling with a common due date. *Journal of Heuristics*, *13*(2), 151–165. https://doi.org/10.1007/s10732-006-9002-2

Min, L., & Cheng, W. (2006). Genetic algorithms for the optimal common due date assignment and the optimal scheduling policy in parallel machine earliness/tardiness scheduling problems. *Robotics and Computer-Integrated Manufacturing*, *22*(4), 279–287. https://doi.org/10.1016/j.rcim.2004.12.005

Mosheiov, G., & Sarig, A. (2009). Due-date assignment on uniform machines. *European Journal of Operational Research*, *193*(1), 49–58. https://doi.org/10.1016/j.ejor.2007.10.043

Nearchou, A. C. (2008). A differential evolution approach for the common due date early/tardy job scheduling problem. *Computers & Operations Research*, *35*(4), 1329–1343. https:// doi.org/10.1016/j.cor.2006.08.013

Sakuraba, S. C., Ronconi, D. P., & Sourd, F. (2009). Scheduling in a two-machine flowshop for the minimization of the mean absolute deviation from a common due date. *Computers & Operations Research*, *36*(1), 60–72. https://doi.org/10.1016/j.cor.2007.07.005

Shabtay, D. (2016). Optimal restricted due date assignment in scheduling. *European Journal of Operational Research*, *252*(1), 79–89. https://doi.org/10.1016/j.ejor.2015.12.043

Shen, X.-N., & Yao, X. (2015). Mathematical modeling and multi-objective evolutionary algorithms applied to dynamic flexible job shop scheduling problems. *Information Sciences*, *298*, 198–224. https://doi.org/10.1016/j.ins.2014.11.036

Vinod, V., & Sridharan, R. (2011). Simulation modeling and analysis of due-date assignment methods and scheduling decision rules in a dynamic job shop production system. *International Journal of Production Economics*, *129*(1), 127–146. https://doi.org/ 10.1016/j.ijpe.2010.08.017

Wang, D.-J., Yin, Y., Cheng, S.-R., Cheng, T. C. E., & Wu, C.-C. (2016). Due date assignment and scheduling on a single machine with two competing agents. *International Journal of Production Research*, *54*(4), 1152–1169. https://doi.org/10.1080/00207543.2015.1056317

Wang, J.-B., Geng, X.-N., Liu, L., Wang, J.-J., & Lu, Y.-Y. (2018). Single machine CON/SLK due date assignment scheduling with controllable processing time and job-dependent learning effects. *The Computer Journal*, *61*(9), 1329–1337. https://doi.org/10.1093/comjnl/bxx121

Xingong, Z., & Yong, W. (2015). Single-machine scheduling CON/SLK due window assignment problems with sum-of-processed times based learning effect. *Applied Mathematics and Computation*, *250*, 628–635. https://doi.org/10.1016/j.amc.2014.11.011

Xiong, X., Wang, D., Edwin Cheng, T. C., Wu, C.-C., & Yin, Y. (2018). Single-machine scheduling and common due date assignment with potential machine disruption. *International Journal of Production Research*, *56*(3), 1345–1360. https://doi.org/10.1080/00207543. 2017.1346317

Yin, N., Kang, L., Sun, T.-C., Yue, C., & Wang, X.-R. (2014). Unrelated parallel machines scheduling with deteriorating jobs and resource dependent processing times. *Applied Mathematical Modelling*, *38*(19–20), 4747–4755. https://doi.org/10.1016/j.apm.2014. 03.022

Yin, Y., Cheng, T. C. E., Cheng, S.-R., & Wu, C.-C. (2013). Single-machine batch delivery scheduling with an assignable common due date and controllable processing times. *Computers & Industrial Engineering*, *65*(4), 652–662. https://doi.org/10.1016/j.cie.2013.05.003

Zhang, R., Song, S., & Wu, C. (2013). A hybrid artificial bee colony algorithm for the job shop scheduling problem. *International Journal of Production Economics*, *141*(1), 167–178. https://doi.org/10.1016/j.ijpe.2012.03.035

Zhao, C., Hsu, C.-J., Lin, W.-C., Liu, S.-C., & Yu, P.-W. (2018). Due date assignment and scheduling with time and positional dependent effects. *Journal of Information and Optimization Sciences*, *39*(8), 1613–1626. https://doi.org/10.1080/02522667.2017.1367515

8 Scheduling with Due-Window Assignment

8.1 INTRODUCTION

Due dates can be determined externally, internally, or jointly with the customer. In case of the due dates determined externally, the firm has to comply with these dates and produce accordingly. Since these due dates do not adhere to the job shop situation, these dates are often not followed, and many jobs finish early or late, which is costly for the firm. For the due dates determined internally or jointly with the customer, the companies can inform the customers about the most suitable due dates for the shop floor conditions.

When external due dates are given, these become parameters that must be followed for the shop floor. In contrast, internal due dates are variables that are tried to be optimized together with scheduling. Flexibility can be provided for the firm's delivery time, and delivery options on suitable due dates and unsuitable due dates put the firm and the shop floor into trouble. On the other hand, late delivery dates do not satisfy the customer, and the customer prefers early delivery dates.

For decades, SWDDA has been a popular study topic. In these studies, scheduling and due-date assignment are done together, and the firm tries to determine the most suitable date for itself. The SWDWA problem is frequently encountered in recent studies. In these problems, the most appropriate time window, a time interval, is determined instead of a single point on the time axis. No penalty is applied to jobs completed within this interval, but jobs completed before and exceeding the time interval are considered early or tardy, and a penalty is applied, Figure 8.1.

Janiak et al. (2015) have comprehensively surveyed the literature concerning problems with various SWDWA models. Firstly, they addressed problems with given due-window or due-windows. Later, they reviewed SWDWA problems with an assignable due window where the due window size is given or assignable due window size and location. Finally, they discussed other due window assignment models. For more information on SWDWA, readers may refer to this survey. Notations commonly used in SWDWA are listed in Tables 8.1 and 8.2 in the following pages.

If jobs are completed before the due window (DW), the opportunity cost of allocating funds, storage, and holding costs may occur. In addition, sometimes, jobs that are completed early can be discarded. When the jobs are completed after the due window, costs such as tardiness penalty, express delivery cost, price reduction, loss of sales, loss of customers, and customer dissatisfaction occur. The firm loses competitiveness, its prestige is damaged, and the customers are unsatisfied with this situation.

Although the long and wide due windows give the company great flexibility, they do not appeal to these customers. This may lead to undesirable results, up to a price

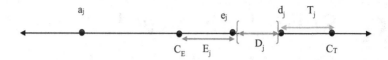

FIGURE 8.1   Scheduling with Due-window.

TABLE 8.1

Notations

d_j^l = Beginning of the due window of job j	a_j = Arrival time of job j
d_j^u = End of the due window of job j	p_j = Total Processing Time of job j
$D_j = d_j^u - d_j^l$ = Due-window of job j	$E_j = \max\{0, d_j^l - C_j\}$ Earliness in case of due window
C_j Completion time (C_E = Early completion, C_T = Tardy Completion)	$T_j = \max\{0, C_j - d_j^u\}$ Tardiness in case of due window

---

reduction, loss of sales, or loss of customers. Late and wide due windows provide little information to customers and make it difficult for customers to make other business plans. Such delivery intervals weaken the manufacturer's competitive edge and reduce customer service performance.

Although early and narrow DWs are very attractive to customers, they put the company in a difficult position. If the DW is too loose, DW cost increases sharply, and if it is too tight, it causes many jobs to be completed early and late, which becomes quite costly. Firms should not give long due intervals, as wide due intervals make it difficult for customers to plan. DWs can be narrowed down by using more resources so as not to adversely affect the prestige of the company, which is costly. In this case, the two variables should be optimized with the scheduling. The most appropriate due-window starting point and optimum width should be decided and integrated with the scheduling decision.

The due window can be applied in many areas, such as semiconductor manufacturing, chemical processing, PERT/CPM scheduling, and IT (Janiak et al., 2015). The SWDWA problem has been studied in single-machine, parallel-machine, and multi-machine environments. In these studies in the literature, the objectives such as the most appropriate scheduling, job sequencing, the most suitable DW beginning, and DW width were tried to be realized. Other objectives like the most suitable batches in batch delivery, the delivery time of each batch, the optimum job compression, and processing times where the processing times of the jobs may vary were tried to be determined. Finally, the jobs assigned to each DW, the most appropriate time for rate modifying activity where maintenance is possible, were tried to be realized. There is a penalty function that the firm tries to minimize while solving these problems. In this penalty function, early jobs, tardy jobs, the starting time of the windows, size of the windows, cost of each delivery, holding costs for the jobs waiting for delivery, the sum of flow times, the number of delivery, processing time

TABLE 8.2
Common Notations Used in the Literature

$I(I_1,...,I_m)$	Set of jobs assigned to each due window	$J = (J_1,...,J_n)$	Set of jobs
		J_{ij}	Job j belongs to group i
$x\,(x_1,...,x_n)$	Set of job compressions	$n\,(n >= 2)$	Number of jobs
a_j, b_j	Compression rate of job J_j	m	Number of batches
$d(d^1{}_1,...,d^1{}_m),d_1$	Starting time of due window(s)	S, π	Job sequence
$d(d^2{}_1,...,d^2{}_m),d_2$	Finish time of due window(s)	$B(B_1,...,B_m)$	Batches
		$G(G_1,...,G_m)$	Groups
$d^1{}_{ij} = p_{ij} + q^1{}_i$	Starting time of due window of job j belongs to group i	$[j]$	Jth position,
		r	rth position
$d^2{}_{ij} = p_{ij} + q^2{}_i$	Finish time of due window of job j belongs to group i	$a_j < 0$ (learning) $a_j > 0$ (aging)	learning/aging factor of job J_j
E_{ij}, T_{ij}	Earliness and tardiness of job j belongs to group i	$D_i = q^2{}_i - q^1{}_i$	DW size for group i
$D(D_1,...,D_m),D$	DWs size ($D = d_2 - d_1$)	C_{max}, C_j	Makespan, Completion time of job j
θ	A constant specified by the decision maker	D_j	Delivery time of job j
ψ	The delivery cost of each batch	$H_j = (D_j - C_j)$ θ	Holding time of job j holding cost
α, α_j	The earliness cost for job J_j	$[B, A]$	The due window for acceptable lead time
β, β_j	The tardiness cost for job J_j	A_j	$A_j = Max\{0, d_j - A\}$ tardiness of due date j
δ, δ_j	1) Unit DW size cost 2) penalty of the earliness of Due-date j	B_j	$B_j = Max\{0, B - d_j\}$ Earliness of due date j
γ, γ_j	1) Unit DW location cost 2) penalty of the Tardiness of Due-date j	$V_j = Earliness$ indicator	(if $C_j < d^1{}_j$ $V_j = 1$, otherwise 0)
v, v_j, G, G_j	Unit RCC for job J_j	$U_j = Tardiness$ indicator	(if $C_j > d^2{}_j$ $U_j = 1$, otherwise 0)
u, u_j	Resource allocated to job J_j	$E_j = Earliness$	$E_j = Max\{d^1{}_j - C_j, 0\}$ when $C_j < d^1{}_j$
w_j, k, v	The workload of job J_j, k and v are positive constants	$T_j = Tardiness$	$T_j = Max\{C_j - d^2{}_j, 0\}$ when $C_j < d^2{}_j$
$bt\,(t \text{ and } b > 0)$	the starting time and the common decreasing rate of processing time	θ_j	Compression rate of job J_j
$\bar{p}_j$	Processing time of job J_j without resource allocating	Q	The upper limit on RCC
p_{jr}	The processing time of job j when assigned to position r	U	The upper limit on scheduling cost
$p^A{}_j = \lambda_j p_j + bs_j$, λ_j, b	Actual processing time at time s_j, Modifying(λ_j) and deteriorating rates (b)	$y_j = 0$ or 1	Represents whether job j is produced or rejected
$\bar{u}_j$	The upper limit on the resource allocation to job J_j	r_j	Rejection cost of job j
$g(r)$	General position function for the rth position ($1 \le r \le n$)	σ	Penalty coefficient of makespan
η	Weight for the total resource consumption cost		

reduction cost, rate modifying activity cost, and for the jobs delivered outside of a time window, are minimized.

The most commonly studied subtopics in the literature on this area are batch delivery, maintenance activity (rate modifying activity), deteriorating/learning maintenance, controllable processing times (allocation of additional resources, the position of the jobs, aging effect, deteriorating effect, learning effect, rate modifying activity) and group scheduling. These topics are the most commonly studied subtopics and are special cases.

8.2 REVIEW OF THE LITERATURE

SWDWA studies are available for different machine environments. Many studies have been conducted for single machine environments (Yin et al., 2013b; Ahmadizar and Farhadi, 2015; Li, 2015; Liu et al., 2017; Mor and Mosheiov, 2017; Wang and Li, 2019). There are also studies for multi-machine and parallel-machine environments (Janiak et al., 2013; Gerstl and Mosheiov, 2013).

SWDWA studies can be classified according to the specific cases in which they are addressed. Batch delivery has been studied in some studies. In many studies, the subject of controllable processing times has been studied, and the process times may vary and decrease according to the additional resources allocated. In some studies, maintenance activities can be used as rate-modifying activities. There are also studies dealing with group scheduling. In addition, deteriorating/learning jobs and deteriorating/learning maintenance were discussed in some studies. Many studies have accepted that linear time-dependent job processing times exist with aging/deteriorating effects.

In studies that considered controllable processing times, process times can constantly change depending on situations such as the allocation of additional resources, the position of the job, the aging effect, the learning effect, the deteriorating effect, and a rate-modifying activity.

Yin et al. (2013b) studied single-machine scheduling with common due window assignment and batch delivery cost. The due window location and size are decision variables, and jobs are delivered in batches. There is no capacity constraint for batches, and batch delivery cost is fixed and independent of the number of jobs. The goal is here to find the best job sequence, delivery date of each job, and due window size and location to minimize penalty function. As a penalty function, the sum of earliness, weighted number of tardy jobs, job holding, due window size and location, and batch delivery costs are tried to be minimized. Mor and Mosheiov (2015) addressed single-machine scheduling and due window assignment with the option of scheduling maintenance (rate modifying) activity. They assumed that the duration of maintenance was a function of its starting time and deteriorating over time. Linear decreasing maintenance time (learning effect) was also considered.

Ahmadizar and Farhadi (2015) studied single-machine scheduling with batch delivery and job release dates. Jobs are released at different points in time but delivered in batches. Every job has a due window. The most appropriate job schedule, batches, and delivery of these batches have been determined to minimize the totals of earliness, tardiness, holding, and delivery costs. The batch size is unlimited, and the delivery cost is fixed and independent of the number of jobs in that batch. Commonly in batch delivery problems, the following assumptions are made: (1) the delivery time for each batch is accepted as the completion time of the last job in that batch, (2) no capacity limit is accepted for the batches, (3) in some studies, the delivery cost for each batch is considered as fixed (4) depending on the number of delivery batches total cost is considered as a non-decreasing function.

Li (2015) addressed single-machine scheduling with due window assignment and batch deliveries where all jobs have a common due window. Batch capacity is assumed unlimited, and the location of the window, start time, and size of the window are decision variables. Optimal job sequence, due window, and delivery times were determined to minimize the total cost of early and tardy delivery, job holding, the due window's start time, the due window, the size of the due window, and the number of deliveries.

Liu et al. (2017) investigated single-machine scheduling with due window assignment, where the processing time of a job depends on its position, start time, and amount of resources allocated to the job. As a penalty function, makespan, earliness, tardiness, due window starting time and size, and the allocated resource cost are tried to be minimized. In their study, it has been tried to determine optimal job schedule, window location, and resource allocation that minimizes total penalty function. Gerstl et al. (2017) studied min-max scheduling with acceptable lead times, where penalties for jobs exceeding pre-specified deadlines determine due dates. In a min-max DIF (Different) model, the value of the objective function is determined by the highest job/due-date cost.

Mor and Mosheiov (2017) addressed single-machine scheduling with the due-window assignment, where they used two competing agents to use a single machine. Every agent has its own objective function, and the goal is to find the joint schedule that minimizes the first objective function subject to an upper bound on the value of the objective function of the second agent. Wang and Li (2019) discussed scheduling problems with a common due window where a due window is negotiable and job processing times depend on resources. They used bicriteria, where the first criterion is a cost function consisting of the weighted numbers of early and late jobs and due window assignment cost. In contrast, the second criterion is total resource consumption cost.

8.3   MATHEMATICAL SOLUTIONS

When the literature is examined, it is seen that different objective functions are used for each case, and the problem is solved according to various performance measures. In each problem, different decisions were made according to the specific situations. The optimum job schedule and sequence were decided, and optimum due windows starting points, and windows sizes, the decision of jobs to be assigned to each due window were determined. Deciding on optimum batches in batch delivery problems, deciding on optimum job compressions, and determining optimum resource consumption were made. The optimal location of RMA (rate modifying activity) in maintenance problems is decided. Finally, the appropriate penalty function was tried to be minimized. Earliness, tardiness, the number of tardy jobs, job holding, windows starting time, windows size, resource consumption cost (RCC), makespan, total batch delivery cost, and job compression costs were minimized inside the penalty function. Before summarizing the literature, it would be appropriate to give the notations commonly used in the literature in Table 8.2.

Yin et al. (2013a) considered single-machine scheduling with batch delivery scheduling and an assignable common due window. They determined the optimal size and location of the common due window. They also decided on optimal batches and dispatch dates for each batch. Finally, they determined the optimal job sequence to minimize the total penalty function, which consists of the sum of earliness, tardiness, job holding, due window, and delivery-related costs of m batches.

$$Z(S, \mathrm{d}_1, \mathrm{D}, \mathbf{B}) = \sum_{J=1}^{n} (\alpha E_j + \beta T_j + \theta H_j + \gamma d_1 + \delta D) + m\psi$$

Yang et al. (2014) studied controllable processing times. They determined optimum starting times and sizes of the windows, optimum batches, optimum job compressions, and optimum schedule to minimize earliness, tardiness, due-windows related costs, and job compression costs.

$$Z(d, D, I, x, \pi) = \sum_{i=1}^{m} \sum_{j \in I_i} (\alpha E_j + \beta T_j + \gamma d_i + \delta D_i + G_j x_j)$$

Li (2015) studied three objective functions. In the first objective function, they minimized total weighed tardy jobs along with the total objective function. The second objective function penalizes an unweighted number of tardy jobs. The cost function Z_3 minimizes the total tardiness penalty inside the objective function.

$$Z_1 = \sum_{J=1}^{n} (\alpha E_j + \beta_j U_j + \theta H_j + \gamma d_1 + \delta D + m\psi)$$

$$Z_2 = \sum_{J=1}^{n} (\alpha E_j + \beta U_j + \theta H_j + \gamma d_1 + \delta D) + m\psi$$

$$Z_3 = \sum_{J=1}^{n} (\alpha E_j + \beta_j T_j + \theta H_j + \gamma d_1 + \delta D) + m\psi$$

Gerstl et al. (2017) studied min-max scheduling with acceptable lead times where the highest job/due-date cost is minimized. Here, they considered position-dependent job processing times and used DIF due dates. Then they extended the model for the due window for acceptable lead times. Here, they assumed a time interval exists where due dates assigned within this interval are not penalized. At the last extension, they allowed job rejection.

P1 $1/\text{DIF}, p_{jr} / \max_{j=1,\dots,n} \left(\alpha E_j + \beta T_j + \gamma A_j \right)$

P2 $1/\text{DWL}, p_{jr} / \max_{j=1,\dots,n} \left(\alpha E_j + \beta T_j + \gamma A_j + \delta B_j \right)$

P3 $1/\text{DIF}, \text{rejection} / \max_{j=1,\dots,n} \left(\alpha E_j + \beta T_j + \gamma A_j \right) y_j + \sum_{j=1}^{n} r_j \left(1 - y_j \right)$

Wang and Li (2019) used two criteria. The first criterion is related to the weighted number of early and tardy jobs and due-window-related costs. The second criterion is related to total resource consumption cost. In this study, S represents the job sequence, and DW start time, size, and resource allocation strategy are determined.

First Criterion: $Z_1(S) = \sum_{j=1}^{n} \left(\alpha_j V_j + \beta_j U_j + \gamma d_1 + \delta D \right)$

Second Criterion: $Z_2(S) = \sum_{j=1}^{n} \left(v_j u_j \right)$

Liu et al. (2017) presented four research problems and solved these problems under CON, SLK, and DIF due-window assignment assumptions. CON due window assigns a common due window for all jobs, and SLK due window assigns an individual due window to each job according to common flow allowance. Finally, DIF due window assigns a different due window to every job with no restrictions. The first two problems assumed linear and convex resource consumption, respectively. The objective is to minimize the sum of earliness, tardiness, due window location, makespan, and resource consumption costs. In the third and fourth problems, they assumed convex resource consumption. In the third problem, they minimized resource consumption costs based on a constraint on total earliness, tardiness, due window location and makespan costs. At the fourth problem, they minimized the sum of earliness, tardiness, due window location, and makespan costs according to resource consumption constraints.

$$Z = \sum_{J=1}^{n} \left(\alpha E_j + \beta T_j + \gamma d_j^1 + \delta D \right) + \sigma \text{Cmax} + \eta \sum_{J=1}^{n} G_j u_j$$

$$\textbf{P1 } 1 \bigg| p_j = \left(\overline{p_j} - bt \right) g(r) - \theta j u j \bigg| \times \sum_{J=1}^{n} \left(\alpha E_j + \beta T_j + \gamma d_j^1 + \delta D \right) + \sigma \text{Cmax} + \eta \sum_{J=1}^{n} G_j u_j$$

$$\textbf{P2 } 1 \bigg| p_j = \left(\left(\frac{w_j}{u_j} \right)^k - bt \right) g(r) \bigg| \times \sum_{J=1}^{n} \left(\alpha E_j + \beta T_j + \gamma d_j^1 + \delta D \right) + \sigma \text{Cmax} + \eta \sum_{J=1}^{n} G_j u_j$$

$$\textbf{P3 } 1 \bigg| p_j = \left(\left(\frac{w_j}{u_j} \right)^k - bt \right) g(r), \sum_{J=1}^{n} \left(\alpha E_j + \beta T_j + \gamma d_j^1 + \delta D \right) + \sigma \text{Cmax} \leq R \bigg| \sum_{J=1}^{n} G_j u_j$$

$$\textbf{P4 } 1 \bigg| p_j = \left(\left(\frac{w_j}{u_j} \right)^k - bt \right) g(r), \sum_{J=1}^{n} u_j \leq U \bigg| \sum_{J=1}^{n} \left(\alpha E_j + \beta T_j + \gamma d_j^1 + \delta D \right) + \sigma \text{Cmax}$$

8.4 CONCLUSION

The SWDDA problem has been widely studied for decades. Although classically, the due dates can be given externally, recently, the due dates have been determined either by the company or jointly by the company and the customer. In the due dates given externally, the company tries to adapt to the given dates and catch up with them. Still, these dates are mostly unaware of the situation of the company and the shop floor, so these dates are often missed, and high penalty costs are incurred. In other cases, more reasonable due dates can be given as the company decides according to its own situation and the situation of the shop floor.

In recent years, instead of due-date, due-window interval assignment problems have become very popular. In these problems, jobs completed in a specific time interval were accepted on time, and no penalty was applied. An early completion penalty was applied to jobs completed before the due window, and a tardiness penalty was applied to jobs completed late. Here, the problem is solved according to the known common due-window in some problems; here, it is considered as the due-window parameter. In some other problems, the common due window is treated as a variable that needs to be determined and integrated with the problem. Here, the starting point of the window and its size have been tried to be determined. Sometimes the starting point is known, but the optimal size is decided. Sometimes the size is known, and the optimal starting point is decided.

In some problems, each job may have its own due window. Sometimes, in case of batch delivery problems, each batch may have a different due window. Special cases in SWDWA problems, batch delivery, maintenance (rate modifying) activity, controllable processing times, group scheduling, aging/deteriorating effects, and learning effects are among the topics that are frequently studied.

REFERENCES

Ahmadizar, F., & Farhadi, S. (2015). Single-machine batch delivery scheduling with job release dates, due windows and earliness, tardiness, holding and delivery costs. *Computers & Operations Research*, 53, 194–205. https://doi.org/10.1016/j.cor.2014.08.012

Gerstl, E., Mor, B., & Mosheiov, G. (2017). Minmax scheduling with acceptable lead-times: Extensions to position-dependent processing times, due-window and job rejection. *Computers & Operations Research*, 83, 150–156. https://doi.org/10.1016/j.cor.2017.02.010

Gerstl, E., & Mosheiov, G. (2013). An improved algorithm for due-window assignment on parallel identical machines with unit-time jobs. *Information Processing Letters*, 113 (19–21), 754–759. https://doi.org/10.1016/j.ipl.2013.06.013

Janiak, A., Janiak, W. A., Krysiak, T., & Kwiatkowski, T. (2015). A survey on scheduling problems with due windows. *European Journal of Operational Research*, 242(2), 347–357. https://doi.org/10.1016/j.ejor.2014.09.043

Janiak, A., Janiak, W., Kovalyov, M. Y., Kozan, E., & Pesch, E. (2013). Parallel machine scheduling and common due window assignment with job independent earliness and tardiness costs. *Information Sciences*, 224, 109–117. https://doi.org/10.1016/j.ins.2012.10.024

Li, C.-L. (2015). Improved algorithms for single-machine common due window assignment and scheduling with batch deliveries. *Theoretical Computer Science*, 570, 30–39. https://doi.org/10.1016/j.tcs.2014.12.021

Liu, L., Wang, J.-J., Liu, F., & Liu, M. (2017). Single machine due window assignment and resource allocation scheduling problems with learning and general positional effects. *Journal of Manufacturing Systems*, 43, 1–14. https://doi.org/10.1016/j.jmsy.2017.01.002

Mor, B., & Mosheiov, G. (2015). Scheduling a deteriorating maintenance activity and due-window assignment. *Computers & Operations Research*, 57, 33–40. https://doi.org/10.1016/j.cor.2014.11.016

Mor, B., & Mosheiov, G. (2017). A two-agent single machine scheduling problem with due-window assignment and a common flow-allowance. *Journal of Combinatorial Optimization*, 33(4), 1454–1468. https://doi.org/10.1007/s10878-016-0049-1

Wang, D., & Li, Z. (2019). Bicriterion scheduling with a negotiable common due window and resource-dependent processing times. *Information Sciences*, 478, 258–274. https://doi.org/10.1016/j.ins.2018.11.023

Yang, D.-L., Lai, C.-J., & Yang, S.-J. (2014). Scheduling problems with multiple due windows assignment and controllable processing times on a single machine. *International Journal of Production Economics*, 150, 96–103. https://doi.org/10.1016/j.ijpe.2013.12.021

Yin, Y., Cheng, T. C. E., Hsu, C.-J., & Wu, C.-C. (2013a). Single-machine batch delivery scheduling with an assignable common due window. *Omega*, 41(2), 216–225. https://doi.org/10.1016/j.omega.2012.06.002

Yin, Y., Cheng, T. C. E., Wang, J., & Wu, C.-C. (2013b). Single-machine common due window assignment and scheduling to minimize the total cost. *Discrete Optimization*, 10(1), 42–53. https://doi.org/10.1016/j.disopt.2012.10.003

9 Integrated Process Planning, Scheduling, and Due-Date Assignment

9.1 INTRODUCTION

Design and manufacturing are essential stages of the product development process. The most critical link between design and production is process planning. Process planning deals with the selection and sequence of production processes necessary to economically and competitively transform a designer's ideas into a physical product (Xu et al., 2011). Process planning is the plan that transforms the raw material into the finished form by considering the design features (Bhaskaran, 1990). Process planning bridges design and manufacturing (Lee and Kim, 2001). Two of the essential activities in a manufacturing system are process planning and scheduling. Computer-aided process planning (CAPP) plays a vital role in the computer-integrated manufacturing (CIM) environment (Kumar and Rajotia, 2003). CAPP is considered the key technology for integrating computer-aided design (CAD) and computer-aided manufacturing (CAM). CAPP uses the computer to assist process planners in planning (Yusof and Latif, 2014).

Scheduling is a decision-making process regularly used in the manufacturing and service industries. It deals with allocating resources to tasks in specific periods and optimizing one or more objectives (Pinedo, 2016). In another definition, scheduling is allocating resources to perform a collection of tasks over time. The purpose of scheduling is to use existing machines effectively, balance the load distribution between different machines, and assign specific tasks to a specific machine. Thus, the efficiency and effectiveness of the shop floor increase (Liao et al., 1994; Kumar and Rajotia, 2003).

Although classically, the due dates were being determined externally, later, the due dates started to be determined internally, jointly with the customer or the company. Determining the due dates that will satisfy the company and the customer will have numerous contributions to the company and customer satisfaction. Since the classically determined due dates were not taking into account the conditions of the shop floor, these dates could be either unreasonably too early or unnecessarily too late, which had many adverse effects on company and customer satisfaction. Recently, internally determined due dates have gained importance, and many studies have been carried out in this area. Now, a crucial managerial issue in production-sales

DOI: 10.1201/9781003215295-9

coordination is the joint determination of order due dates between the customer and the manufacturer through the sales personnel (Lawrence, 1994).

When the literature is examined, many studies have been carried out in the field of integrated process planning and scheduling (IPPS) in recent decades. Likewise, a great deal of work has been done in the field of scheduling with due-date assignment (SWDDA) over the past few decades. Although IPPS and SWDDA problems are trendy research topics, integrated process planning, scheduling, and due-date assignment (IPPSDDA) is relatively new. To the best of our knowledge, there are only a few recent studies (Demir et al., 2015; Erden et al., 2021) on this subject.

Process planning, scheduling, and due-date assignment are three important manufacturing functions. Out of these functions, the output of the process planning becomes the input for the scheduling. Due dates received or given, unaware of the process plans and scheduling will not be realistic and will often not be followed on the shop floor. Three essential manufacturing functions are integrated quite widely in pairs but not in threes, except few studies (Erden et al., 2019, 2021; Demír et al., 2021).

The first sub-integration problem in these studies is the IPPS problem. Here, the process planning function is integrated with the scheduling function, but unfortunately, the due date has not been optimized together (Yu et al., 2018; Barzanji et al., 2020). Traditionally, process planning and scheduling were handled separately and sequentially. These two functions, which were solved differently, were locally optimized independently, away from the global optimum. The output of the process planning was the input for the downstream scheduling function. A bad input had a negative impact on the downstream scheduling function and, consequently, the performance of the shop floor. For more detailed information on IPPS studies, see Chapter 5.

In process planning, which was prepared without knowing the scheduling, process planners could often choose the machines they preferred from the shop floor. While these mostly chosen machines became the bottlenecks, the other machines they did not prefer remained idle by a large percentage. This attitude caused a significant decrease in shop floor performance, and there could be large load imbalances between machines in the shop. Traditionally, a single process plan was prepared for each product instead of several alternative process plans.

Developments in software and hardware have led to developments in the field of CAPP, as in every other subject (Liao et al., 1994; Kumar and Rajotia, 2003; Xu et al., 2011; Yusof and Latif, 2014). Developments in the field of CAPP have led to the preparation of process plans with better quality, ease, and lower cost, and also facilitated the preparation of alternative process plans. Process plans, which are now integrated with scheduling and being aware of the shop floor condition, provide better quality input for downstream scheduling. When there is an undesirable situation on the shop floor or when there is a load imbalance in the machines, thanks to the prepared alternative process plans, the jobs can be directed to other machines, and alternative process plans are activated. For more detailed information on process planning and CAPP, see Chapter 2.

Integrated process planning and scheduling provide better overall performance for these reasons. Therefore, it is useful to integrate these two functions. Classically, there were no alternative process plans in the scheduling independent and sequential planning approach. Process planners can often choose the machines they like while

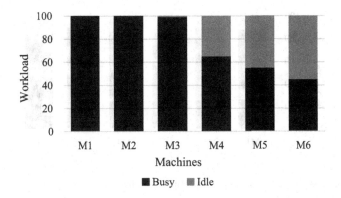

FIGURE 9.1 Possible machine load in traditional process planning and scheduling.

making their plans so that some machines may experience congestion and some machines may remain idle at a high rate, as seen in Figure 9.1, since there is only one process plan. There is no integration between process planning and scheduling, and there is a sequential relationship between them, as in Figure 9.2.

When process planning and scheduling are integrated, a bidirectional relationship occurs between them, as in Figure 9.3. Process planners prepare process plans suitable for the shop floor conditions and also prepare alternative plans that schedulers can choose when necessary. Thus, better quality input is provided for downstream scheduling, and alternative plans are prepared that the shop floor can choose when necessary.

In this case, since the shop floor conditions are considered, and due to the integration of process planning and scheduling, as well as the availability of alternative

FIGURE 9.2 Relationship in traditional process planning and scheduling.

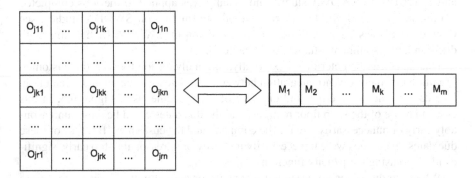

FIGURE 9.3 Relationship in integrated process planning and scheduling.

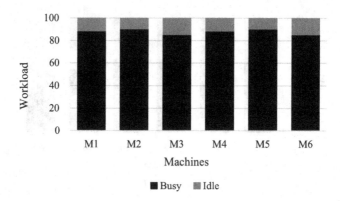

FIGURE 9.4   Possible machine load in integrated process planning and scheduling.

process plans when necessary, it is possible to achieve a more balanced machine load, as in Figure 9.4.

where

M_i : i^{th} machine M_m : There are m machines on the shop floor
O_{jk} : k^{th} operation of jth job O_{jn} : j^{th} job has a maximum of n operations
O_{jkk}: k^{th} operation of kth route of jth job O_{jrn}: job j has a maximum of r routes and n operations

SWDDA, another sub-integration problem, has been widely studied in the last decades, and many articles have been published (Geng et al., 2019). In these studies, scheduling and due dates were tried to be determined in an integrated manner, but unfortunately, process planning was not included in integration and optimization. Jobs completed before the due dates were considered early and penalized. Jobs completed after the due dates were considered tardy and penalized similarly. Contrary to many studies in the literature Erden et al. (2019) and Erden et al. (2021) also penalized the length of the due dates. The logic here is the reality that no customer will prefer long due dates. In relatively more recent studies in this area, a due-windows have been tried to be assigned instead of the due dates Janiak et al. (2015) and Wang and Li (2019). In SWDWA studies, no penalty was applied to the jobs completed within the due windows. For more detailed information on SWDDA studies, see Chapter 7; likewise, for more detailed information on SWDWA (Scheduling with due-window assignment) studies, see Chapter 8.

Due dates can be determined externally, internally, or jointly with the customer. Traditionally, there was no integration between due-date assignment and scheduling. Due dates could be external, and unaware of the schedule, as in Figure 9.5. In this case, unaware of the shop floor realities, and the due dates could be either unreasonably early or unnecessarily late for the company and the customers. Because of these due dates, many jobs were unnecessarily too tardy or unreasonably too early, significantly increasing the penalty function.

When the due dates are assigned internally or agreed with the customer, more realistic and more successful due dates can be given since these dates are given

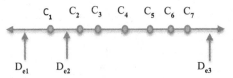

FIGURE 9.5 External Due-date assignment.

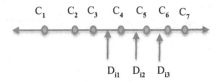

FIGURE 9.6 Internal Due-date Assignment.

according to the shop floor conditions, as in Figure 9.6. In this case, unreasonable excessive earliness or unnecessary excessive tardiness problems are eliminated, the penalty function value decreases, and the overall performance can increase significantly.

where

C_i = Completion time of ith job
D_{ei} = External due-date assignment (One of possible (e) predetermined due dates)
D_{ii} = Internal due date assignment (One of possible (i) internally determined due dates)

Traditionally, process planning, scheduling, and due-date assignment were done sequentially and separately, as in Figure 9.7. When a process plan is prepared, process planners often prefer the machine they like and often do not choose the machines they do not like in the process plans. This separation would cause a load imbalance in the machines and cause a decrease in efficiency. Some machines were constantly busy, while others were often idle. Process plans were often not followed when things got stuck and went wrong on the shop floor. In addition, the due dates were not determined in accordance with the shop floor conditions or determined

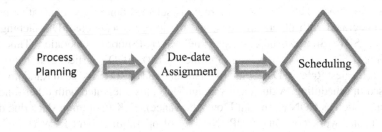

FIGURE 9.7 Traditional process planning, scheduling, and due-date assignment.

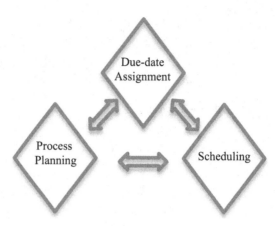

FIGURE 9.8   Integrated process planning, scheduling, and due-date assignment.

internally. Thus, either unreasonably too early due dates would be assigned, and the shop floor would not be able to deliver on the promise or too late due dates would be given, and many early jobs would be unnecessarily penalized. Since these three functions are not integrated and not optimized jointly, a very high penalty function will result, and the overall performance will be poor.

However, as shown in Figure 9.8, the overall performance will improve substantially when the three functions are integrated. Process plans will be prepared according to the conditions of the shop floor, and alternative process plans will be prepared to have alternative routes when necessary, providing more balanced machine loads on the shop floor. In case of congestion, jobs can be directed to different machines according to alternative process plans. According to the state of the shop floor and process plans, the due dates given internally or jointly with the customer will be more realistic, and unnecessary earliness and tardiness penalties will be minimized. Unreasonably early dates will not be given, and promises that cannot be kept will not be made. Moreover, unnecessarily long due dates will not be given, the prestige and competitiveness of the company will not be damaged, and customer dissatisfaction will not occur. The result will significantly improve overall performance, and the penalty function will be reduced considerably.

Process planning is easier to prepare with CAPP after developments in hardware and software. Alternative process plans prepared can be helpful in case of congestion on the shop floor. The most suitable route is selected among the alternative process plans integrated with other functions, and the jobs are scheduled. Dispatching rules such as SPT (Shortest Processing Time), SOT (Shortest Operation Time), MS (Minimum Slack), ATC (Apparent Tardiness Cost), EDD (Earliest Due date), ERD (FIFO) (Earliest Release Date, First in First Out), and SIRO (Service in Random Order) are used in scheduling. A due date is assigned to jobs integrated with other functions, and rules such as CON (Constant Flow Allowance), SLK (Common slack due dates), TWK (Total work content), NOP (Number-of-operations), and PPW (Processing-plus-wait) are used when assigning. Due-date assignment rules are explained in detail in Chapter 4. Dispatching rules are mentioned in Chapter 3. For this reason, in

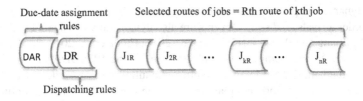

FIGURE 9.9   Chromosome representation of IPPSDDA problem.

this chapter, the rules for determining the due dates and the priority rules in scheduling will not be mentioned in detail.

Demir and Erden (2017) and Demir et al. (2021) used the chromosome structure in their IPPSDDA studies. The first two genes in the chromosomes are dominant genes and represent the due-date assignment rules and the dispatching rules used. As in Figure 9.9, other genes show which route was or will be chosen for each job.

9.2   REVIEW OF THE LITERATURE

If we discuss the literature in order, starting with the sub-integrations, it would be appropriate to discuss IPPS studies first. In addition, since it is explained in detail in the IPPS Chapter 5, too much detail will not be given here. One of the first studies on this topic (Meenakshi Sundaram and Fu, 1988) argued that integrating process planning and scheduling would lead to increased productivity. Later, Bhaskaran (1990) discussed selecting the most appropriate process plans among the alternatives to minimize the total process time, the total number of process steps, the resources used, and reduce the manufacturing costs.

In some studies, the IPPS problem was divided into loading and scheduling subproblems and solved sequentially. In the loading section, the routes were selected with mathematical models. In the scheduling problem, the jobs were assigned to the machines according to the designated routes with desired dispatching rules, and their sequence and schedule were decided (Hutchison et al., 1991; Demir and Wu, 1996). Apart from these studies, other studies divide the complex problem into several models, parts, modules, and phases (Yu et al., 2015; Barzanji et al., 2020). Over time, some review studies have been made on IPPS. Some of these studies can be listed as follows: Li et al. (2010); Phanden et al. (2011). For detailed information, readers may refer to these survey papers. It is better to see detailed information on IPPS (Phanden et al., 2020).

Z. Zhang et al. (2016) studied energy-conscious integrated process planning and scheduling problem. X. X. Li et al. (2015) optimized multi-objectives, including energy consumption, makespan, and balanced machine utilization. Barzanji et al. (2020) solved the IPPS problem by developing a logic-based Bender's decomposition (LBBD) algorithm. With the LBBD algorithm, the decision variables in the IPPS problem were divided into two models: the master problem and the sub-problem. The master problem determines the process plan and operation-machine assignment, while the sub-problem optimizes sequencing and scheduling decisions.

Some IPPS and IPPSDDA studies have been considered for a static shop floor environment (Demir et al., 2021). Some IPPS and IPPSDDA studies have been considered for the dynamic shop floor environment and machine breakdowns, new order arrivals, urgent order arrivals, and similar events are also dynamically considered in these shop floors (L. Zhang and Wong, 2015; Xia et al., 2016; Erden et al., 2019, 2021).

The SWDDA problem, another sub-problem of the IPPSDDA problem, is also widely studied. In the SWDDA problem, scheduling and due-date assignment are treated together, but the process planning function is not included in the optimization and integration here (Gordon et al., 2012).

Most research in the SWDDA field has been done for the single-machine environment (Kacem and Kellerer, 2019). T'kindt and Della Croce (2012) can be given as an example of two machine environments. As an example of a flow shop environment, readers may refer to Geng et al. (2019). For examples of parallel machine environments, readers may examine Liu et al. (2013). Finally, Baykasoğlu and Göçken (2009) and Vinod and Sridharan (2011) are examples of job shop–type machining environments.

9.3 METAHEURISTIC SOLUTIONS

Jin et al. (2015) proposed a hybrid honeybee mating optimization (HBMO) algorithm combining the HBMO algorithm and variable neighborhood search (VNS) to solve the IPPS problem. Dai et al. (2015) studied energy-conscious integrated process planning and scheduling for job shops. Due to the complexity of the problem, they used a modified genetic algorithm to discover the optimal solution (Pareto solution) between energy consumption and makespan. M. Yu et al. (2015) proposed a hybrid algorithm based on genetic algorithm (GA) and particle swarm optimization (PSO) for optimizing the IPPS problem. L. Zhang and Wong (2015) proposed an object-coding genetic algorithm (OCGA) to resolve the IPPS problems in a job shop type of flexible manufacturing system.

X. Liu et al. (2016) solved the mathematical model they proposed for the IPPS problem with the ACO (ant colony optimization) algorithm. X. X. Li et al. (2015) simulated honeybee mating and annealing processes to solve the IPPS problem. Xia et al. (2016) studied the dynamic IPPS problem. They dynamically accounted for machine breakdowns, urgent order arrivals, and similar events. They used a hybrid genetic algorithm with a variable neighborhood search (GAVNS) method to solve the problem due to its good search performance.

Sobeyko and Mönch (2017) discussed an integrated process planning and scheduling problem in large-scale flexible job shops (FJSs). Because of the complex nature of the investigated problem, an iterative scheme was designed that is based on variable neighborhood search (VNS) on the process planning level. M. R. Yu et al. (2018) proposed a discrete particle swarm optimization (DPSO) algorithm to solve the IPPS optimization problem.

Erden et al. (2019) handled the problem of IPPSDDA in a dynamic environment with many metaheuristic methods in their studies and solved the problem with GA, SA (Simulated Annealing), and TA (Tabu Algorithm) methods and a hybrid form of SA and TA methods with the GA metaheuristic. Erden et al. (2021) studied the IPPSDA problem in a dynamic environment and solved the problem with some PSO variants and genetic algorithm metaheuristics. Demir et al. (2021) compared simulated annealing and evolutionary strategies and their hybrid methods with random search in another study and solved the IPPSDDA problem in a static environment.

9.4 CONCLUSION

Although IPPS and SWDDA problems have been widely studied in recent decades, the IPPSDDA problem has been covered in very few studies and is a relatively new research topic. Although the first studies on this subject were carried out in a static shop floor environment, the problem has been solved in recent studies according to the dynamic shop floor environment and constantly updated incoming orders.

Before the IPPS and SWDDA problems, three essential manufacturing functions were handled separately, considered, and optimized locally. These three functions' conflicting goals and solution methods were detrimental to global performance. Traditionally, a single process plan was prepared, and process planners did not always consider the shop floor when performing their functions. Many times, they often preferred some high-performance machines or did not prefer other machines. This selection would create an overload imbalance between the machines and severely affect the performance and efficiency of the shop floor. In backlogs and unexpected situations, these plans could not be followed. Because process plans are input for shop floor scheduling, bad inputs negatively impact scheduling and reduce shop floor performance and efficiency. Recently, process plans and even alternative plans that can be used when necessary that take into account the situation of the shop floor have started to be prepared. Developments in the hardware, software, and algorithms have worked very well in process planning, as in every other subject. It has become easier to prepare a process plan with the help of computers. Thus, preparing alternative plans became easier, economical, and high quality.

Traditionally, due dates were not integrated with scheduling, especially with process plans, and were often determined externally. These dates would be either unreasonably too early or unnecessarily too late. This awkwardness would result in unnecessary earliness and tardiness costs. In addition, although only tardiness was punished classically, earliness was also punished as it was not wanted as a result of the spread of the JIT philosophy. In addition, the internal due-date assignment approach started to be applied instead of the external due-date assignment. In the internal due-date assignment, both the customer and the firm decide on the due dates jointly, or the company gives the due dates in accordance with its own conditions. These due dates are more realistic, and no unreasonably too early or too late dates are given.

Studies in which the three functions are integrated are relatively new and few. Only a few studies are available, especially for the dynamic shop floor environment. In these studies, order arrivals were considered dynamic. In the future, other events, such as machine breakdowns, can be considered and would be a new subject of study.

REFERENCES

Barzanji, R., Naderi, B., & Begen, M. A. (2020). Decomposition algorithms for the integrated process planning and scheduling problem. *Omega, 93*, 102025. https://doi.org/10.1016/j.omega.2019.01.003

Baykasoğlu, A., & Göçken, M. (2009). Gene expression programming based due date assignment in a simulated job shop. *Expert Systems with Applications*, *36*(10), 12143–12150

Bhaskaran, K. (1990). Process plan selection. *International Journal of Production Research*, *28*(8), 1527–1539. https://doi.org/10.1080/00207549008942810

Dai, M., Tang, D., Xu, Y., & Li, W. (2015). Energy-aware integrated process planning and scheduling for job shops. *Proceedings of the Institution of Mechanical Engineers, Part B: Journal of Engineering Manufacture*, *229*(1_suppl), 13–26. https://doi.org/10.1177/0954405414553069

Demir, H. I., Cakar, T., Ipek, M., Uygun, O., & Sari, M. (2015). Process planning and due-date assignment with ATC dispatching where earliness, tardiness and due-dates are punished. *Journal of Industrial and Intelligent Information*, *3*(3). 197–204. https://doi.org/10.12720/jiii.3.3.197-204

Demir, H. I., & Erden, C. (2017). Solving process planning and weighted scheduling with WNOPPT weighted due-date assignment problem using some pure and hybrid metaheuristics. *SAU Journal of Science*, *21*(2), 210–222. https://doi.org/10.16984/saufenbilder.297014

Demir, H. I., Phanden, R., Kökçam, A., Erkayman, B., & Erden, C. (2021). Hybrid evolutionary strategy and simulated annealing algorithms for integrated process planning, scheduling and due-date assignment problem. *Academic Platform Journal of Engineering and Science*, *9*(1), 86–91. https://doi.org/10.21541/apjes.764150

Demir, H. I., & Wu, S. D. (1996). *A comparison of several optimization schemes for the integrated process planning and production scheduling problems* [Master of Science Thesis]. Lehigh University.

Erden, C., Demir, H. I., & Canpolat, O. (2021). A modified integer and categorical PSO algorithm for solving integrated process planning, dynamic scheduling and due date assignment problem. *Scientia Iranica*. https://doi.org/10.24200/sci.2021.55250.4130

Erden, C., Demir, H. I., & Kökçam, A. H. (2019). Solving integrated process planning, dynamic scheduling, and due date assignment using metaheuristic algorithms. *Mathematical Problems in Engineering*, *2019*, 1–19. https://doi.org/10.1155/2019/1572614

Geng, X.-N., Wang, J.-B., & Bai, D. (2019). Common due date assignment scheduling for a no-wait flowshop with convex resource allocation and learning effect. *Engineering Optimization*, *51*(8), 1301–1323. https://doi.org/10.1080/0305215X.2018.1521397

Gordon, V., Strusevich, V., & Dolgui, A. (2012). Scheduling with due date assignment under special conditions on job processing. *Journal of Scheduling*, *15*(4), 447–456. https://doi.org/10.1007/s10951-011-0240-2

Hutchison, J., Leong, K., Snyder, D., & Ward, P. (1991). Scheduling approaches for random job shop flexible manufacturing systems. *International Journal of Production Research*, *29*(5), 1053–1067. https://doi.org/10.1080/00207549108930119

Janiak, A., Janiak, W. A., Krysiak, T., & Kwiatkowski, T. (2015). A survey on scheduling problems with due windows. *European Journal of Operational Research*, *242*(2), 347–357. https://doi.org/10.1016/j.ejor.2014.09.043

Jin, L., Zhang, C., & Shao, X. (2015). An effective hybrid honey bee mating optimization algorithm for integrated process planning and scheduling problems. *The International Journal of Advanced Manufacturing Technology*, *80*(5–8), 1253–1264. https://doi.org/10.1007/s00170-015-7069-3

Kacem, I., & Kellerer, H. (2019). Complexity results for common due date scheduling problems with interval data and minmax regret criterion. *Discrete Applied Mathematics*, *264*, 76–89. https://doi.org/10.1016/j.dam.2018.09.026

Kumar, M., & Rajotia, S. (2003). Integration of scheduling with computer aided process planning. *Journal of Materials Processing Technology*, *138*(1–3), 297–300. https://doi.org/10.1016/S0924-0136(03)00088-8

Lawrence, S. R. (1994). Negotiating due-dates between customers and producers. *International Journal of Production Economics*, *37*(1), 127–138. https://doi.org/10.1016/0925-5273(94)90013-2

Lee, H., & Kim, S.-S. (2001). Integration of process planning and scheduling using simulation based genetic algorithms. *The International Journal of Advanced Manufacturing Technology, 18*(8), 586–590. https://doi.org/10.1007/s001700170035

Li, X., Gao, L., Zhang, C., & Shao, X. (2010). A review on integrated process planning and scheduling. *International Journal of Manufacturing Research, 5*(2), 161–180.

Li, X. X., Li, W. D., Cai, X. T., & He, F. Z. (2015). A hybrid optimization approach for sustainable process planning and scheduling. *Integrated Computer-Aided Engineering, 22*(4), 311–326.

Liao, T. W., Coates, E. R., Aghazadeh, F., Mann, L., & Guha, N. (1994). Modification of CAPP systems for CAPP/scheduling integration. *Computers & Industrial Engineering, 25*(1–4), 203–206.

Liu, M., Zheng, F., Wang, S., & Xu, Y. (2013). Approximation algorithms for parallel machine scheduling with linear deterioration. *Theoretical Computer Science, 497*, 108–111. https://doi.org/10.1016/j.tcs.2012.01.020

Liu, X., Ni, Z., & Qiu, X. (2016). Application of ant colony optimization algorithm in integrated process planning and scheduling. *The International Journal of Advanced Manufacturing Technology, 84*(1–4), 393–404. https://doi.org/10.1007/s00170-015-8145-4

Meenakshi Sundaram, R., & Fu, S. (1988). Process planning and scheduling—A method of integration for productivity improvement. *Computers & Industrial Engineering, 15* (1–4), 296–301. https://doi.org/10.1016/0360-8352(88)90102-7

Phanden, R. K., Jain, A., & Davim, J. P. (Eds.) (2020). *Integration of Process Planning and Scheduling: Approaches and Algorithms*. CRC Press.

Phanden, R. K., Jain, A., & Verma, R. (2011). Integration of process planning and scheduling: A state-of-the-art review. *International Journal of Computer Integrated Manufacturing, 24*(6), 517–534. https://doi.org/10.1080/0951192X.2011.562543

Pinedo, M. L. (2016). *Scheduling*. Springer International Publishing. https://doi.org/10.1007/978-3-319-26580-3.

Sobeyko, O., & Mönch, L. (2017). Integrated process planning and scheduling for large-scale flexible job shops using metaheuristics. *International Journal of Production Research, 55*(2), 392–409. https://doi.org/10.1080/00207543.2016.1182227

T'kindt, V., & Della Croce, F. (2012). A note on "Two-machine flow-shop scheduling with rejection" and its link with flow-shop scheduling and common due date assignment. *Computers & Operations Research, 39*(12), 3244–3246. https://doi.org/10.1016/j.cor. 2012.04.009

Vinod, V., & Sridharan, R. (2011). Simulation modeling and analysis of due-date assignment methods and scheduling decision rules in a dynamic job shop production system. *International Journal of Production Economics, 129*(1), 127–146. https://doi.org/10.1016/ j.ijpe.2010.08.017

Wang, D., & Li, Z. (2019). Bicriterion scheduling with a negotiable common due window and resource-dependent processing times. *Information Sciences, 478*, 258–274. https://doi. org/10.1016/j.ins.2018.11.023

Xia, H., Li, X., & Gao, L. (2016). A hybrid genetic algorithm with variable neighborhood search for dynamic integrated process planning and scheduling. *Computers & Industrial Engineering, 102*, 99–112. https://doi.org/10.1016/j.cie.2016.10.015

Xu, X., Wang, L., & Newman, S. T. (2011). Computer-aided process planning – A critical review of recent developments and future trends. *International Journal of Computer Integrated Manufacturing, 24*(1), 1–31. https://doi.org/10.1080/0951192X.2010.518632

Yu, M. R., Yang, B., & Chen, Y. (2018). Dynamic integration of process planning and scheduling using a discrete particle swarm optimization algorithm. *Advances in Production Engineering & Management, 13*(3), 279–296. https://doi.org/10.14743/apem2018.3.290

Yu, M., Zhang, Y., Chen, K., & Zhang, D. (2015). Integration of process planning and scheduling using a hybrid GA/PSO algorithm. *The International Journal of Advanced Manufacturing Technology, 78*(1–4), 583–592. https://doi.org/10.1007/s00170-014-6669-7

Yusof, Y., & Latif, K. (2014). Survey on computer-aided process planning. *The International Journal of Advanced Manufacturing Technology*, *75*(1–4), 77–89. https://doi.org/10.1007/s00170-014-6073-3

Zhang, L., & Wong, T. N. (2015). An object-coding genetic algorithm for integrated process planning and scheduling. *European Journal of Operational Research*, *244*(2), 434–444. https://doi.org/10.1016/j.ejor.2015.01.032

Zhang, Z., Tang, R., Peng, T., Tao, L., & Jia, S. (2016). A method for minimizing the energy consumption of machining system: Integration of process planning and scheduling. *Journal of Cleaner Production*, *137*, 1647–1662. https://doi.org/10.1016/j.jclepro.2016.03.101

10 Dynamic Integrated Process Planning, Scheduling, and Due-Date Assignment

10.1 INTRODUCTION: BACKGROUND AND DRIVING FORCES

In a classical job shop scheduling problem (JSSP), each job has separate routes, and a schedule of jobs is prepared such that the operations are processed on the machines in the manufacturing system. The solution to the scheduling problem has several objectives, such as the efficient use of resources, timely and fast response to orders, and minimizing early or late completion according to due dates (Baker & Trietsch, 2009). JSSP is divided into different categories under various assumptions and configurations, such as static, dynamic, and stochastic scheduling. In the static model, the following assumptions were made (Hollier, 1975; Holthaus & Ziegler, 1997):

- All jobs are available at time zero.
- Jobs consist of several operations.
- Each job is sent directly to the machine for its first operation.
- Each job was independent of all other jobs.
- Each job may have to wait in the machine queue.
- Each machine can only operate one operation at the same time.
- Each job can be performed on one machine at the same time.
- Jobs have deterministic processing times (including setup times) and due dates.
- Machine breakdowns or rush orders were not allowed.

Dynamic scheduling is one of the current areas of study related to the job-shop scheduling problem. Dynamic scheduling problems address situations frequently encountered in real job shops, such as unexpected arrival over time, machines breaking down, rushing or canceling orders, and changes in processing time. The dynamic scheduling problem may be deterministic or stochastic, depending on the ready time of the jobs. If the scheduling options (arrival times, due dates, routes, and machine statuses of jobs) are known in advance, scheduling is deterministic. Stochastic scheduling occurs if the times are distributed randomly according to a specific distribution (Lin et al., 1997). The various definitions for dynamic scheduling studies can be defined as follows:

- **Dynamic scheduling**: Events such as continuously arriving new jobs and machine breakdowns disrupt the current scheduling.

DOI: 10.1201/9781003215295-10

- **Rescheduling**: When a dynamic event occurs, static scheduling must be resolved from scratch under new circumstances.
- **Event-driven rescheduling**: Events are divided into those that need to be rescheduled and those that do not. Similar to the rescheduling strategy, a new schedule for every job is generated for the first category of occurrences. No response was displayed for the second batch of events until the next rescheduling point.
- **Partial rescheduling**: This does not start from scratch to create a schedule for every job. When a dynamic event occurs in a system, it attempts to modify a portion of the schedule.
- **Performance-driven rescheduling**: This compares the predicted performance from the created schedule to the actual performance metric. The rescheduling approach is used if the difference between these metrics is greater than a predetermined threshold.
- **Offline schedule**: produced for every open job at once throughout the horizon.
- **Online schedule**: Scheduling decisions are made one at a time as needed to reflect shifting system conditions.

It is necessary to model and solve scheduling problems in real job shops in a dynamic environment. The schedules prepared in the job shops are reviewed when new jobs arrive, and the scheduling process is repeated. Therefore, the scheduling should be dynamic. In this chapter, the studies carried out in a dynamic environment are included, and an example is provided with a small application.

10.2 REVIEW OF THE LITERATURE

Considerable research studies on dynamic job-shop scheduling have been conducted. Numerous studies have also been conducted in early literature reviews (Allahverdi et al., 1999, 2008; Allahverdi & Aydilek, 2010; Ramasesh, 1990). Adibi et al. (2010) used a neighboring solution algorithm to solve dynamic events, such as unexpected job arrivals and machine breakdowns. Dominic et al. (2004) developed two new dispatching rules for dynamic job-shop scheduling. It has been shown that referral rules perform better than existing rules, such as SPT, LIFO, and LPT. Aydin and Öztemel (2000) solved the dynamic job shop scheduling problem using reinforced learning factors in order to select appropriate dispatching rules. Li et al. (2005) developed an artificial neural network using a genetic algorithm to solve this problem. Sha and Liu (2005) developed a data-mining model that can adapt to dynamic conditions in a dynamic job shop environment. Zandieh and Adibi (2010) used artificial neural networks to estimate the appropriate scheduling method parameters for the shortest average processing time. Zhang et al. (2013) developed a hybrid tabu search algorithm for a dynamic, flexible job shop scheduling problem.

Studies on dynamic integrated process planning and scheduling problems are limited and are usually based on agent-based or metaheuristic solutions. Chun and Wong (2003) solved the DIPPS problem using a hybrid multi-factor-based system. Lin et al. (2001) integrated process planning and scheduling problems to avoid the dynamic events that may occur in a job shop. Xia et al. (2016) solved the problem of machine failure and random arrival of jobs using a neighbor search algorithm and the benefit of alternative process plans.

10.3 PROBLEM FORMULATIONS

In scheduling problems, there are a limited number of jobs n and a limited number of machines m (Pinedo, 2008a). Scheduling problems consist of three elements. Suppose these elements are shown as $\alpha < \beta < \gamma$; α symbolizes machines and has a single input. β defines details about process characteristics and constraints, and may contain no input, or it may have a single entry, or it may have multiple inputs. γ defines the objective function to minimize the problem and usually contains a single input (Pinedo, 2008b). The objective of the scheduling problem is to minimize the completion time of each job or the average number of tardy jobs. This study determined the objective function of tardiness, earliness, and optimization of the given due date. In cases of earliness, the enterprise must consider several costs. The tardiness on jth job (T_j) and earliness of jth job (E_j) in the study are calculated as shown in Equations 10.1 and 10.2:

$$T_j = \max\left(c_j - d_j, 0\right) \tag{10.1}$$

$$E_j = \max\left(d_j - c_j, 0\right) \tag{10.2}$$

There is tardiness if the job is completed after a given due date. In the case of tardy jobs, the penalty for early completion is 0. If the job is completed before the due date, there is early completion, and the tardiness penalty is 0. The penalty values of or early completion (P_E) and late completion (P_T) and due date (P_D) are calculated as shown in the equations below. Equation (6) shows the total penalty. The objective function $(f_{\min})$ of the model is defined as:

$$P_D = w_j \times \left(8 \times \left(\frac{D}{480}\right)\right) \tag{10.3}$$

$$P_E = w_j \times \left(5 + 4 \times \left(\frac{E}{480}\right)\right) \tag{10.4}$$

$$P_T = w_j \times \left(6 + 6 \times \left(\frac{T}{480}\right)\right) \tag{10.5}$$

$$P_j = P_D + P_E + P_T \tag{10.6}$$

$$f_{\min} = \sum_{j=1}^{n} P_j \tag{10.7}$$

If the tardiness is positive, the jth job is completed late. If it is negative, then the jth job is completed early.

The assumptions made in this study are as follows:

- The arrival times were random, with an exponential distribution.
- The processing time of each operation is non-deterministic and normally distributed.
- Each job had its process route.
- If an operation needs to be performed before an operation, it cannot be performed without it.
- The machines do not contain breakdowns.

10.3.1 Dispatching Rules

When a machine is idle, it is necessary to prioritize all pending jobs to select the next job and determine the job to be processed on the machine. For the assignment, two features of all pending jobs can be considered: characteristics of the jobs (weights, processing time, or due dates) and machines (the speed of the machine, the number of jobs waiting to be processed, and the total duration of pending operations). Priority values were calculated for all the remaining jobs, and the job with the highest index value was selected for the process. Among the dispatch rules, the most basic are:

- Apparent Tardiness Cost (ATC)
- Earliest due date first (EDD)
- Earliest release date first (ERD)
- First In First Out(FIFO), Last In First Out (LIFO)
- Longest Operation Time (LOT)
- Longest Processing Time (LPT)
- Minimum Slack (MS)
- Service in random order (SIRO)
- Shortest Operation Time (SOT)
- Shortest Processing Time (SPT)

The dispatching rules can be divided into static and dynamic rules. Static dispatching rules are not time-bound. These rules vary only according to the jobs or machine data. In addition, the dynamic rules were time-dependent. In these rules, calculations that vary over time are made. Thus far, different dispatching rules have been used in the literature for scheduling problems, and the advantages of referral rules have been stated. Several comparative studies have evaluated the performance of these referral rules. No definitive study has been conducted on which dispatching rule determines the global best value of the scheduling problem. Some dispatching rules provide outstanding results in some scheduling problems and poor results in some problems. The efficient operation of the dispatching rules varies according to the structure of the problem, process constraints, and the objective function of the production.

However, many studies have concluded that the best solutions are obtained by changing the dispatching rules according to dynamic conditions and becoming suitable for the system (Chiang & Fu, 2007; Đurasević & Jakobović, 2018, 2019; Gohareh & Mansouri, 2022; Jun et al., 2019; Oukil & El-Bouri, 2021; Pergher & de Almeida, 2018; Vlašić et al., 2019).

Mouelhi-Chibani and Pierreval (2010) developed an artificial neural network that determines the best dispatching rule according to the structure of the problem. The study showed that the artificial neural network could automatically determine an efficient dispatching rule.

Kaban et al. (2012) compared the dispatching rules using simulations. In this study, 44 hybrid or single dispatching rules were evaluated. As a result of the study, it was stated that MTWR (Most Total Work Remaining) dispatching rule achieved the best result. In general, hybrid dispatching rules provide more successful results than single rules. We selected a jobshop-type production area from the automotive industry in this study. The study also noted how critical dispatching rules are related to global scheduling performance.

10.3.2 DUE DATE ASSIGNMENT RULES

The problem of machine scheduling related to predetermined due dates has been studied previously. However, assigning a due date is a decision-making process of great importance regarding production efficiency. The advantages of due-date assignment functions that are performed in conjunction with job scheduling have been demonstrated in many studies (Kanet, 1981; Szwarc, 1989). In all these studies, scheduling was performed by providing a standard due date for the jobs. However, as a requirement for lean production and the growing need for flexible production, jobs often have different due dates. Table 10.1 explains the due date rules used in this chapter.

TABLE 10.1
Due Date Assignment Rules

Rule	Formula
Slack (SLK)	$d_i = a_i + p_i + q_x$
Weighted slack (WSLK)	$d_i = a_i + p_i + w_{1x}q_x$
Total work content (TWK)	$d_i = a_i + k_x p_i$
Weighted total work content (WTWK)	$d_i = a_i + w_{1x}k_x p_i$
Number of operations plus processing time (NOPPT)	$d_i = a_i + p_i + 5k_x o_i$
Weighted number of operations plus processing time (WNOPPT)	$d_i = a_i + p_i + 5w_{1x}k_x o_i$
Random-allowance due dates (RDM)	$d_i = a_i + N\sim(3P_{av}, P_{av})$
Processing-time-plus-wait (PPW)	$d_i = a_i + k_x p_i + q_x$
Weighted processing-time-plus-wait (WPPW)	$d_i = a_i + w_{1x}k_x p_i + w_{2x}q_x$

d_i: Due date, a_i: Arrival time, p_i: Processing time, p_av: Average processing time.

10.3.3   ARRIVAL TIMES OF THE JOBS

Each job enters the system according to an exponential distribution with an average of $1/\mu$. The formula used for the exponential distribution is as follows:

$$a_j = \frac{1}{\mu} = \frac{\mu_p \mu_g}{UM} \tag{10.8}$$

μ: frequency of arrival of jobs, μ_p: average processing time per operation, μ_g: average number of operations for each job, U: job shop utilization rate, M: number of machines in the job shop

The probability distribution function (PDF) and the cumulative distribution function (CDF) of the exponential distribution are given as follows:

$$f(x) = \begin{cases} \lambda e^{-\lambda x} & x \geq 0 \\ 0 & x < 0 \end{cases} \tag{10.9}$$

$$F(x) = \begin{cases} 0 & x < 0 \\ 1 - e^{-\lambda x} & x \geq 0 \end{cases} \tag{10.10}$$

10.4   A SAMPLE DIPPSDDA STUDY

Two outputs are obtained from the solution of the DIPPSDDA system to be integrated. The first is the process plans produced separately for each job, and the second is the schedule suitable for the objective function, which shows which work will be processed on which machines, when, and to what extent. Therefore, there is a problem with choosing and assigning. If solutions are found for these two subproblems, a suitable DIPPSDDA system will be created.

Dynamic scheduling primarily operates as static scheduling. When a new job arrives at the job shop, the job list is updated, and the problem becomes dynamic scheduling. In this study, a dynamic scheduling problem was investigated. Moreover, discrete event simulations were created separately for each job shop to verify the dispatch performance and due-date rules. To obtain more effective solutions for the simulations, it was assumed that the exponential distribution distributed the arrival times of the jobs, and the normal distribution distributed the operation times.

In short, the job has different routes, and it completes the production process by visiting the machines in the job shop in order on the route. When a new job arrives, it enters the machine's process queue according to the operations of the jobs according to the route selected by the developed algorithm. Machines are chosen based on the dispatching rules generated by the algorithm from the pending operations. After the due dates of the jobs were calculated, the simulation was run, and the completion times for each job were obtained as the output from the simulation. The simulation was run until all the jobs were completed.

The steps to run the simulation were as follows:

- Run the algorithm steps and create an individual solution.
- When a new job arrives, it determines the weights, processing times, operation priority and times, and the job route.

- The due date is calculated based on the job selected due date assignment rule.
- The first job operation entered the machine queue to be assigned.
- The machine selects the operation from the pending operations based on the dispatching rule.
- The completion time for the final operation of the job was determined by the departure time of the work.
- The objective function calculates each job's earliness or tardiness and due dates.
- The solution is optimized using an algorithm.

10.5 CASE STUDY

A mini-job shop was created for the case study because of its easy follow-up and calculation. There are four machines and two jobs in the mini-job shop. The processing times are given as averages, and a normal distribution with standard deviation produces times. The generated input data are presented in Table 10.2.

For the simulation study, animation was developed using the Salabim package, a simulation package in Python programming language. The simulation is shown in Figure 10.1, which depicts the animation interface. First, a mini-shop floor was produced and used for the animation. The visual aspects of the events on the mini-shop floor are shown based on the developed program. The queues of the machines in the job shop and the queues between the operations of the jobs are shown in the figure. According to the dispatching rule, the machine selects from among the pending operations. The jobs' operations, operation times, and route information of the jobs are also shown in the figure. Priority relationships exist between job operations. Accordingly, the other job operations cannot be processed before the previous operation is completed. Red operations are pending, whereas green operations are traded. In the upper-right corner of the animation, the operation times can be followed with the help of a clock. After an operation is completed, it is removed from the animation, and the animation is run until all operations are completed.

The step-by-step simulation results at the end of the animation are presented in Table 10.3.

TABLE 10.2
Operation Times and Route Information of the Jobs Produced for the Mini-job Shop

Job No	Route No	Operation Times				Machines				Weight	Arrival Time
0	0	10	9	23	17	0	1	1	1	1	16
	1	19	2	27	1	0	1	0	0		
1	0	1	12	45	7	1	0	1	1	0.33	20
	1	1	8	10	13	0	1	1	0		
2	0	7	23	3	4	1	1	1	0	0.33	22
	1	23	17	5	13	0	0	1	1		
3	0	10	8	7	10	1	0	1	0	1	45
	1	23	4	13	6	0	1	1	1		

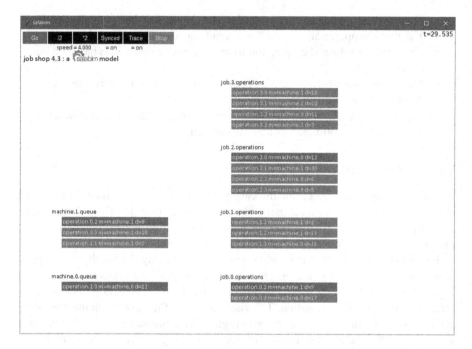

FIGURE 10.1   Visualization of simulation work.

TABLE 10.3
Scheduling Results

Chromosome None Fitness 26.074416666666664 Route [0, 0, 1, 0]
Due Date Rule 11 Dispatching Rule 8
job 0 arrival_time= 16 departure_time= 90 due-date= 175.29999999999998 weight= 1.0
operation.0.0 m=machine0 d=10 start_time= 16 finish_time= 26
operation.0.1 m=machine1 d=9 start_time= 26 finish_time= 35
operation.0.2 m=machine1 d=23 start_time= 35 finish_time= 58
operation.0.3 m=machine1 d=17 start_time= 73 finish_time= 90
job 1 arrival_time= 20 departure_time= 149 due-date= 551.8181818181818 weight= 0.33
operation.1.0 m=machine1 d=1 start_time= 20 finish_time= 21
operation.1.1 m=machine0 d=12 start_time= 66 finish_time= 78
operation.1.2 m=machine1 d=45 start_time= 97 finish_time= 142
operation.1.3 m=machine1 d=7 start_time= 142 finish_time= 149
job 2 arrival_time= 22 departure_time= 162 due-date= 496.54545454545456 weight= 0.33
operation.2.0 m=machine0 d=23 start_time= 26 finish_time= 49
operation.2.1 m=machine0 d=17 start_time= 49 finish_time= 66
operation.2.2 m=machine1 d=5 start_time= 68 finish_time= 73
operation.2.3 m=machine1 d=13 start_time= 149 finish_time= 162
job 3 arrival_time= 45 departure_time= 107 due-date= 139.5 weight= 1.0
operation.3.0 m=machine1 d=10 start_time= 58 finish_time= 68
operation.3.1 m=machine0 d=8 start_time= 78 finish_time= 86
operation.3.2 m=machine1 d=7 start_time= 90 finish_time= 97
operation.3.3 m=machine0 d=10 start_time= 97 finish_time= 107

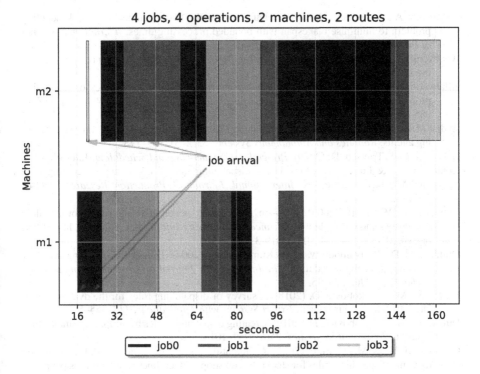

FIGURE 10.2   Gantt chart obtained as a result of the case study.

In the resulting solution produced by the developed algorithm, the solution [11, 8, 0, 0, 1, 1] was evaluated, the performance value was calculated as 26.07, and the Gantt diagram of the solution is shown in Figure 10.2.

10.6   CONCLUSION

Given an original process plan and timing at the outset, the job shop executes according to the schedule until problems occur. Scheduling will continue until there is an unexpected situation, selecting the appropriate manufacturing resources and sorting out their operations. However, in today's job shops, there are problems that we can describe as unexpected, unwanted, or disruptive. Such situations may include machine failure, rush orders that must be included in the scheduling, etc. To solve such situations, it is imperative that scheduling be developed in a way that adapts to conditions that cannot be predicted in advance and need to be solved urgently. Owing to the development of a system that can adjust itself according to the negative situations that may occur, inefficient situations that may arise during the scheduling of jobs and the determination of delivery dates can be eliminated, and scheduling targets can be achieved.

REFERENCES

Adibi, M. A., Zandieh, M., & Amiri, M. (2010). Multi-objective scheduling of dynamic job shop using variable neighborhood search. *Expert Systems with Applications*, *37*(1), 282–287.

Allahverdi, A., & Aydilek, H. (2010). Heuristics for the two-machine flowshop scheduling problem to minimise makespan with bounded processing times. *International Journal of Production Research*, 48(21), 6367–6385.

Allahverdi, A., Gupta, J. N., & Aldowaisan, T. (1999). A review of scheduling research involving setup considerations. *Omega*, 27(2), 219–239.

Allahverdi, A., Ng, C. T., Cheng, T. E., & Kovalyov, M. Y. (2008). A survey of scheduling problems with setup times or costs. *European Journal of Operational Research*, 187(3), 985–1032.

Aydin, M. E., & Öztemel, E. (2000). Dynamic job-shop scheduling using reinforcement learning agents. *Robotics and Autonomous Systems*, 33(2–3), 169–178.

Baker, K. R., & Trietsch, D. (2009). *Principles of Sequencing and Scheduling*. John Wiley.

Chiang, T.-C., & Fu, L.-C. (2007). Using dispatching rules for job shop scheduling with due date-based objectives. *International Journal of Production Research*, 45(14), 3245–3262.

Chun, A., & Wong, R. (2003). N*—An agent-based negotiation algorithm for dynamic scheduling and rescheduling. *Advanced Engineering Informatics*, 17, 1–22. https://doi.org/10.1016/S1474-0346(03)00019-3

Dominic, P. D., Kaliyamoorthy, S., & Kumar, M. S. (2004). Efficient dispatching rules for dynamic job shop scheduling. *The International Journal of Advanced Manufacturing Technology*, 24(1), 70–75.

Đurasević, M., & Jakobović, D. (2018). A survey of dispatching rules for the dynamic unrelated machines environment. *Expert Systems with Applications*, 113, 555–569.

Đurasević, M., & Jakobovićć, D. (2019). Creating dispatching rules by simple ensemble combination. *Journal of Heuristics*, 25(6), 959–1013.

Gohareh, M. M., & Mansouri, E. (2022). A simulation-optimization framework for generating dynamic dispatching rules for stochastic job shop with earliness and tardiness penalties. *Computers & Operations Research*, 140, 105650.

Hollier, R. H. (1975). A review of: Introduction to sequencing and scheduling. *International Journal of Production Research*, 13(6), 654–654. https://doi.org/10.1080/00207547508943038

Holthaus, O., & Ziegler, H. (1997). Improving job shop performance by coordinating dispatching rules. *International Journal of Production Research*, 35(2), 539–549. https://doi.org/10.1080/002075497195894

Jun, S., Lee, S., & Chun, H. (2019). Learning dispatching rules using random forest in flexible job shop scheduling problems. *International Journal of Production Research*, 57(10), 3290–3310.

Kaban, A. K., Othman, Z., & Rohmah, D. S. (2012). Comparison of dispatching rules in job-shop scheduling problem using simulation: A case study. *International Journal of Simulation Modelling*, 11(3), 129–140. https://doi.org/10.2507/IJSIMM11(3)2.201

Kanet, J. J. (1981). Minimizing the average deviation of job completion times about a common due date. *Naval Research Logistics Quarterly*, 28(4), 643–651.

Li, Y., Liu, Y., & Liu, X. (2005). Active vibration control of a modular robot combining a back-propagation neural network with a genetic algorithm. *Journal of Vibration and Control*, 11(1), 3–17.

Lin, C.-W. R., Chen, H.-Y., & Xiao, Q.-S. (2001). Dynamic integrated process planning and scheduling. *Journal of the Chinese Institute of Industrial Engineers*, 18(2), 21–32. https://doi.org/10.1080/10170660109509169

Lin, S.-C., Goodman, E. D., & Punch III, W. F. (1997). A genetic algorithm approach to dynamic job shop scheduling problem. Proceedings of the 7th International Conference on Genetic Algorithm, 481–488, San Francisco, CA, USA.

Mouelhi-Chibani, W., & Pierreval, H. (2010). Training a neural network to select dispatching rules in real time. *Computers & Industrial Engineering*, 58(2), 249–256.

Oukil, A., & El-Bouri, A. (2021). Ranking dispatching rules in multi-objective dynamic flow shop scheduling: A multi-faceted perspective. *International Journal of Production Research, 59*(2), 388–411.

Pergher, I., & de Almeida, A. T. (2018). A multi-attribute, rank-dependent utility model for selecting dispatching rules. *Journal of Manufacturing Systems, 46*, 264–271.

Pinedo, M. (2008a). *Scheduling: Theory, Algorithms, and Systems*, Springer.

Pinedo, M. L. (2008b). Scheduling. In *Scheduling*. https://doi.org/10.1007/978-0-387-78935-4.

Ramasesh, R. (1990). Dynamic job shop scheduling: A survey of simulation research. *Omega, 18*(1), 43–57.

Sha, D. Y., & Liu, C.-H. (2005). Using data mining for due date assignment in a dynamic job shop environment. *The International Journal of Advanced Manufacturing Technology, 25*(11), 1164–1174.

Szwarc, W. (1989). Single-machine scheduling to minimize absolute deviation of completion times from a common due date. *Naval Research Logistics (NRL), 36*(5), 663–673.

Vlašić, I., Đurasević, M., & Jakobović, D. (2019). Improving genetic algorithm performance by population initialisation with dispatching rules. *Computers & Industrial Engineering, 137*, 106030.

Xia, H., Li, X., & Gao, L. (2016). A hybrid genetic algorithm with variable neighborhood search for dynamic integrated process planning and scheduling. *Computers & Industrial Engineering, 102*. https://doi.org/10.1016/j.cie.2016.10.015

Zandieh, M., & Adibi, M. A. (2010). Dynamic job shop scheduling using variable neighbourhood search. *International Journal of Production Research, 48*(8), 2449–2458. https://doi.org/10.1080/00207540802662896

Zhang, L., Gao, L., & Li, X. (2013). A hybrid genetic algorithm and tabu search for a multi-objective dynamic job shop scheduling problem. *International Journal of Production Research, 51*(12), 3516–3531. https://doi.org/10.1080/00207543.2012.751509

11 Solution Techniques in Integrated Manufacturing Functions

11.1 INTRODUCTION

Scheduling tasks and processes is a crucial aspect of many industries, and finding efficient solutions to these problems can significantly impact productivity and profitability. Different manufacturing functions are related to each other. Integrating them and solving this problem as a holistic approach is more efficient. Individual solutions cannot improve the whole system as they have limited information about conflicting objectives or unexpected events in the manufacturing system.

Although the due-date assignment single-machine scheduling problem can be solved in polynomial-time (Wang and Guo, 2010), process planning and scheduling problems are both in the NP (Non-deterministic Polynomial-time)—Hard problem class and cannot be solved in polynomial-time. Integration of these functions results in even more complex problems.

There are several approaches to handling integrations in these problems. For example, the IPPS problem can be separated into two subproblems. The first is called master, which is used to address the high-level subproblem, and the second is called slave, which is used for dealing with the low-level subproblem. These two subproblems are handled with different approaches, as given in Maravelias and Sung (2009) and Zhang et al. (2022), which are hierarchical, iterative, and full-space.

In another approach, the IPPS problem can be divided into three categories, as given in Baykasoglu and Ozbakir (2009), Li et al. (2010), and Phanden et al. (2019), which are non-linear, closed-loop, and distributed approaches. The other integrations can also be regarded with these categories.

Examples of the integration of different manufacturing functions are given in Table 11.1.

In the Web of Science core collection following search query is written. In this search query, an asterisk (*) is a wildcard, and it represents any group of characters, including no character. For example, "integrat*" can stand for "integrated," "integration," or "integrating," etc. Moreover, parentheses and Boolean operators ("and," "or") are also used to refine the search process. Here "TS" stands for topic search, which includes the title, abstract, author keywords, and Keywords Plus sections of publications.

Only studies in the English language are taken into consideration which counts to 790 between the years 1990 and 2023. Data obtained were analyzed using the Biblioshiny interface of the Bibliometrix library in R language.

It is found that there is an almost 11% publication growth rate in this area. Mean citations per article and mean citations per year are given with the number of

DOI: 10.1201/9781003215295-11

TABLE 11.1

Some of the Examples of Integration of Manufacturing Functions

Function 1	Function 2	Function 3	Function 4	Literature Example
scheduling	maintenance planning			Sharifi and Taghipour (2021)
scheduling	predictive maintenance			Zhai et al. (2021)
distributed production	distribution			Fu et al. (2022)
production scheduling	transportation scheduling			Li et al. (2022)
scheduling	distribution operations			Pereira and Nagano (2022)
economic lot and delivery scheduling	closed-loop supply chain with open-shop manufacturing			Goli and Davoodi (2018)
customer relationship management	production planning and control			Gucdemir and Selim (2017)
scheduling	dispatching	routing of jobs and AGVs		Umar et al. (2015)
scheduling	outbound distribution scheduling	due date assignment		Assarzadegan and Rasti-Barzoki (2016)
scheduling	process planning	due date assignment		Erden et al. (2019)
scheduling	material requirement planning	production planning	transportation planning	Shafiee-Gol et al. (2021)

publications as given in Figure 11.1. While the red line indicates the mean of total citations per article in the related year, the green area shows the mean of total citations divided by the publication age. Thus, it can be inferred that studies with integrated manufacturing functions still have decent growth potential.

Keywords in these studies are investigated and sorted with usage number, genetic algorithm, multi-agent system, mixed integer programming, simulation, hybrid algorithm, particle swarm optimization, ant colony optimization, simulated annealing, and tabu search methods are mostly utilized for solutions.

Based on these findings, we will first examine exact methods which provide a precise solution to the problem at hand but may be computationally expensive. Next we will discuss heuristic methods which offer a suboptimal solution but are much faster to compute. Metaheuristic methods which involve a higher-level search process will also be covered. We will then delve into agent-based methods which utilize autonomous agents to find solutions and hybrid methods which combine multiple solution techniques. Finally, we will examine the use of machine learning methods in solving integrated manufacturing functions. This chapter aims to overview the various solution techniques applied in this context comprehensively.

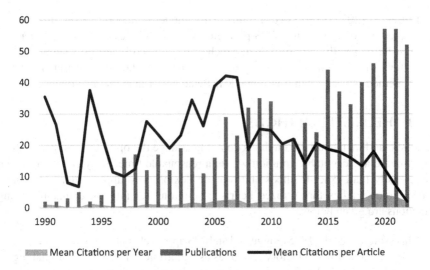

FIGURE 11.1 Annual publication numbers with mean citations in integrated manufacturing topics.

11.2 EXACT METHODS

Exact methods ensure that the optimum solution is found by traversing the whole search space. Obtaining the optimum solution is quite appealing. However, it requires significant computing power. That is nearly impossible to solve most of the problems. It can only be utilized for rather small-sized problems.

Although many mathematical and mixed-integer models exist for solving job shop scheduling problems deterministically, there are few studies on integrated problems, as this problem is more complicated. Some of them are given below.

Shi et al. (2014) presented a method for solving the mixed-integer dynamic optimization problem involved in the DIPPS and dynamic optimization in large-scale production processes, which decomposes the problem into mixed-integer linear programming (MILP) form. They compared their method with simultaneous approaches and found that it reduces computing time by more than three orders of magnitude.

Barzanji et al. (2020) developed a logic-based Benders decomposition algorithm to find exact solutions for IPPS problems with two relaxations and two enhancements to improve performance and achieve faster convergence.

Gao et al. (2021) developed an approach for concurrent optimization of process planning, job-shop scheduling, and open-field layout in a reconfigurable manufacturing system to improve sustainability by reducing production costs, hazardous waste, and greenhouse gas emissions while also reducing product tardiness.

They proposed a multi-objective mathematical model and used both brute-force search and non-dominated sorting genetic algorithm (NSGA) III to find Pareto-optimal solutions.

Liu et al. (2022a) proposed a multi-MILP model collaborative optimization method for solving the IPPS problem, which combines deterministic exact searching with an approximate stochastic framework and uses submodels for exact searching under a greedy framework. They demonstrated their method is faster than traditional MILP and better than metaheuristics.

11.3   HEURISTIC METHODS

Heuristic and metaheuristic methods do not guarantee the optimum results, but they can provide an acceptable solution within a reasonable time. They can be designed more easily than exact methods and provide flexibility to apply different problem types. Some studies that utilized heuristic solutions for integrated manufacturing functions are given below.

Chryssolouris et al. (1985) suggested an IPPS approach that considers assigning production tasks to factory resources as a multiple-criteria decision-making problem. They developed a decision-making approach for the manufacturing environment that can minimize or eliminate problems caused by conflicting objectives or unexpected changes.

Huang et al. (1995) used a progressive approach to address IPPS by starting at a global level and working towards a more detailed level, which reduces computational complexity and can be implemented in time-sensitive manufacturing environments.

Ausaf et al. (2015) suggested the priority-based heuristic algorithm that integrates dispatching rules with prioritizing jobs to optimize the IPPS problem and improve system performance. Their approach effectively achieves superior results for complex problems while using fewer computational resources in testing.

11.4   METAHEURISTIC METHODS

Metaheuristic methods are a class of optimization algorithms that can solve complex optimization problems with multiple conflicting objectives that are uncertain or change over time by guiding the search process of other heuristic methods. These methods are mostly inspired by natural phenomena in physics, chemistry, or biology. They are designed to be flexible and adaptable, allowing them to solve numerous problems. Here we investigate five of them, mostly utilized in the solutions for integrated manufacturing functions.

11.4.1   Genetic Algorithms (GA)

GA is a metaheuristic optimization algorithm inspired by natural selection and genetics principles to find solutions to optimization problems. It was first introduced by John Holland in the 1970s and has since been widely applied to various optimization problems. The GA generates a population of candidate solutions, each representing a possible schedule for the set of jobs. These candidate solutions are then evaluated based on their quality, with the best solutions being selected to form the next generation of candidates. This process continues until an optimal schedule is found or a predetermined number of iterations has been reached (Holland, 1992).

GA is the most used algorithm in various problems within integrated manufacturing functions. Some studies that utilized GA to solve integrated manufacturing functions are given below.

Lee and Kim (2001) proposed an approach to the IPPS using simulation-based GA, which reduces makespan and tardiness through the computation of performance measures based on process plan combinations and their incorporation into GA.

Lin et al. (2001) suggested a DIPPS system that uses a two-stage, GA-based control mechanism to dynamically generate optimal process plans and production schedules in response to disturbances. They evaluated their system through statistical optimization techniques and found it significantly improved total production cost in simulation.

Lihong and Shengping (2012) presented a mathematical model for IPPS and an improved GA for optimizing them simultaneously, including new initial selection, genetic representation, and operator methods. Their model significantly reduces the makespan and produces satisfactory results for the mean flow time.

Xia et al. (2016) presented a DIPPS model and a hybrid GA with variable neighborhood search to solve it. They tested machine breakdown and new job arrival disturbances and were found to significantly improve upon traditional methods for DIPPS.

Zhang et al. (2020) proposed a hierarchical multi-strategy GA to minimize energy consumption, makespan, and peak power in a job shop through IPPS. Their approach finds a set of non-dominated solutions more effective than the NSGA II approach when evaluated using the C-metric, spacing metric, and maximum spread values. They conducted a case study, resulting in a 15% reduction in energy consumption for two different scheduling schemes with the same makespan.

Li and Gao (2020) introduced an integration model and a modified GA-based approach for IPPS, including more efficient genetic representations and operator schemes, and successfully tested for superiority and adaptablity compared to other methods.

Liu et al. (2021) proposed novel integrated encoding and decoding methods and a modified GA based on them for solving the IPPS problem, successfully tested on 37 well-known open problems. Their approach outperformed the comparative algorithms in a real-world case from a nonstandard equipment workshop.

11.4.2 Ant Colony Optimization

The ant colony optimization (ACO) method is a metaheuristic optimization algorithm inspired by ants' behavior in their search for food. It involves the creation of a population of virtual ants, which search for solutions to the optimization problem by leaving a trail of pheromones as they move. The strength of the pheromone trail is influenced by the quality of the solutions found by the ants, and the ants can use this information to guide their search for better solutions (Dorigo et al., 1996).

Some studies that utilized ACO solutions for integrated manufacturing functions are given below.

Leung et al. (2010) suggested an ACO algorithm in an agent-based system to IPPS, using artificial ants implemented as software agents in a multi-agent system platform and a graph-based solution method to minimize makespan. Their simulation studies showed that the ACO could solve IPPS problems effectively, and the agent-based approach can provide a distributive algorithm computation.

Rossi and Lanzetta (2020) proposed a rule-oriented system that extracts features from CAD, a process planner that identifies alternative modes and allows swaps between operations, and a framework that produces a precedence graph of manufacturing features with flexibility via both conjunctive and disjunctive arcs and nodes, which can be visited by ACO and reduces the node redundancy of AND/OR graphs while dealing with the increased complexity of IPPS.

11.4.3 Particle Swarm Optimization (PSO)

The PSO method is a metaheuristic optimization algorithm inspired by the behavior of swarms of animals, such as birds or fish. It involves the creation of a population of particles, which move through the solution space and are influenced by the movements of their neighbors. The particles can adapt their movements based on the quality of the solutions they have encountered, allowing them to search for good solutions to the optimization problem (Kennedy and Eberhart, 1995).

Some studies that utilized PSO solutions for integrated manufacturing functions are given below.

Guo et al. (2009) presented a modified version of the PSO algorithm to solve the IPPS problem. They tested the modified PSO algorithm through case studies and compared it to the GA and SA algorithms, showing that it can generate satisfactory results.

Yu et al. (2018) proposed a two-phase approach for DIPPS and a discrete PSO algorithm to solve it, utilizing an external archive and mutation operation to improve local search. They successfully tested for static and dynamic situations.

11.4.4 Simulated Annealing (SA)

SA is a heuristic optimization method used to search for a function's global minimum, which was introduced in 1983 (Kirkpatrick et al., 1983). It is named after the annealing process used in metallurgy, which purifies and softens metals by heating them to a high temperature and then gradually cooling them. The basic idea behind SA is to start with a random initial solution and then gradually improve it by making small, random changes. The changes are accepted or rejected based on the value of the function at the new solution, with the probability of acceptance being greater for solutions that improve the function and smaller for solutions that worsen it. This process is repeated until the function reaches a minimum or until a predetermined stopping criterion is met.

Some studies that utilized SA solutions for integrated manufacturing functions are given below.

Li and McMahon (2007) suggested a unified model and a SA based method for the IPPS in a job shop that handles a lot of make-to-order business, using three strategies for exploring the search space and various performance criteria to meet practical requirements.

Rostami et al. (2020) investigated integrated scheduling of production and distribution activities in a supply chain with considerations for machine deterioration and learning effects where the manufacturer aims to minimize total weighted completion time. The distributor aims to reduce shipping times by batch delivery using capacitated vehicles. They modeled the problem as a MILP with a branch and bound algorithm for a special case and a SA algorithm for large-scale instances.

Shafiee-Gol et al. (2021) presented an integrated Mixed-Integer Nonlinear Programming model that addresses the dynamic decision-making of parts scheduling, material requirement planning, production planning, and transportation planning for designing a cellular manufacturing system in a three-layer supply chain by considering market demands, heterogeneous vehicles, raw materials, machine capacity, transportation time and cost, operation time, alternative processing routes and dynamic cell formation. They solved it with an improved SA algorithm with a matrix-based chromosome representation and a sequential procedure for generating initial solutions.

11.4.5 Tabu Search (TS)

TS is used to find approximate solutions to combinatorial optimization problems introduced in 1986 (Glover, 1986). It is a local search-based method, starting with an initial solution and then iteratively making small changes to improve the solution. The key idea behind TS is to prevent the algorithm from getting stuck in a local optimum by using a tabu list, which is a memory structure that keeps track of the solutions that have been recently visited. Solutions on the tabu list are avoided in the search process to encourage the algorithm to explore new regions of the search space.

Some studies that utilized TS solutions for integrated manufacturing functions are given below.

Wang et al. (2019) proposed an integrated single-machine scheduling and multi-vehicle routing problem under the carbon emission policy, which minimizes the total carbon emissions in manufacturing and distribution by considering the possibility of switching machines in a period between two adjacent workpieces. They solved it using a mathematical programming model and a TS hybrid algorithm. They extended it to minimize total costs and minimize both total costs and carbon emissions, to guide green manufacturing and logistics for industrial enterprises.

Li et al. (2022) addressed the integrated production and transportation scheduling problem in hybrid flow shops, incorporating transportation scheduling on Automated Guided Vehicles and using an effective genetic TS algorithm to minimize makespan.

11.5   AGENT-BASED METHODS

Agent-based methods are a class of computational approaches that simulate the actions and interactions of multiple agents within a system. These agents can be designed to represent real-world entities such as people, animals, or machines. They can be programmed to exhibit certain behaviors and make decisions based on their surroundings and goals. Some studies that utilized agent-based solutions for integrated manufacturing functions are given below.

Shukla et al. (2007) suggested a bidding-based multi-agent system for the IPPS problem, in which tool cost is treated as a dynamic quantity and predicted using a data-mining agent with a fuzzy decision tree. When a job arrives, machine agents bid on features, and the optimal or near-optimal process plans and schedules are found using a hybrid TS-SA algorithm.

Nejad et al. (2008) developed a multi-agent architecture for a DIPPS, utilizing a negotiation protocol and heuristic search algorithms to generate more appropriate process plans and schedules in a dynamically changing environment.

Nejad et al. (2011) developed a multi-agent architecture for a DIPPS in manufacturing systems, utilizing a negotiation protocol based on alternative manufacturing processes and coordination agents for preparing the process plans and schedules more efficiently in a dynamically changing environment and cope with disturbances. They tested their model through simulation software and integrated it with the open robot interface network architecture for potential real-world application.

Zhang et al. (2012) proposed a multi-agent system architecture for solving the DIPPS problem with embedded heuristic algorithms, which can be combined with various heuristic methods for DIPPS, resulting in a system with high flexibility, extensibility, and accessibility for manufacturing applications.

11.6   HYBRID METHODS

Hybrid solution methods combine two or more approaches to solve complex optimization problems. An example of a hybrid method is the combination of mathematical programming and heuristics. This approach uses a mathematical model to optimize the solution, while heuristics guide the search for a solution and improve its efficiency. Another example of a hybrid method is the combination of metaheuristics and ML. This approach uses metaheuristics to explore the solution space and identify promising solutions, while ML algorithms are used to fine-tune and improve the solutions. Hybrid methods can also involve the combination of multiple metaheuristics.

In this approach, different metaheuristics are used in combination to explore the solution space and improve the quality of the solution.

Some studies that utilized hybrid solutions for integrated manufacturing functions are given below.

Chan et al. (2006) proposed an Artificial Immune System-based method coupled with a fuzzy logic controller as a solution for an IPPS model that incorporates an outsourcing strategy to optimize performance in a rapidly changing environment.

Zhao et al. (2006) utilized a fuzzy inference system and PSO to select appropriate machines for IPPS based on the machine's reliability characteristics and balancing the load for each machine.

Wong et al. (2006) proposed a hybrid multi-agent system for IPPS in flexible manufacturing environments, using decentralized negotiations with a comprehensive set of flexibilities and a supervisory agent to coordinate and improve global performance, and they successfully tested on machine breakdown and new part arrival disruptions in large-scale rescheduling problems.

Amin-Naseri and Afshari (2012) suggested a hybrid GA for solving the IPPS in a modern manufacturing system, using problem-specific genetic operators, local search, and a novel neighborhood function considering flexible job shop constraints and nonlinear precedence relations to find optimal or near-optimal solutions.

Pereira and Nagano (2022) discussed the importance of integrating and synchronizing manufacturing and logistics activities in today's complex business environment. They presented different heuristic methods based on iterated greedy, iterated local search, SA, and variable neighborhood search for scheduling production and distribution operations in no-wait flow shop systems integrated with transportation systems that can work with distinct distribution plans. Their aim is to improve operational performance and increase customer satisfaction.

Liu et al. (2022b) investigated the Multi-Objective Distributed IPPS problem, which aims minimizing the makespan, maximum machine load, and total machine load in distributed manufacturing patterns. They proposed a MILP model and Multi-Objective Memetic Algorithm that utilizes OR-nodes encoding, genetic operators, and SA mechanism to optimize solutions while considering precedence constraints of the processes.

By integrating the supplier portfolio into production scheduling with a customer-imposed delivery time limit, Mahmud et al. (2022) addressed the issues of offering highly customized and on-time delivery needs at the lowest cost. They modeled using the flexible job shop problem and solved it with two metaheuristic algorithms based on multi-objective PSO, including a TS-inspired search mechanism, a modified mutation operator, two easy to implement approaches to address inadequate archive diversity, and leader selection strategies.

11.7   MACHINE LEARNING (ML) METHODS

ML methods are a type of artificial intelligence that can solve a wide range of problems in manufacturing. Below are some studies that utilized ML solutions for integrated manufacturing functions.

Park et al. (1993) investigated the use of explanation-based learning to improve intelligent process planning by combining the "variant" and "generative" approaches. They implemented a learning process planner called explanation-based learning for intelligent process planning (EXBLIPP) and demonstrated that it has many of the intended advantages but also suffers from brittleness that limits its ability to respond to unpredictable features of the environment. To address this problem, the authors used contingent explanation-based learning to defer certain planning decisions until execution time, allowing EXBLIPP to adapt to the dynamic environment of a manufacturing system. Their paper discusses this approach's strengths and weaknesses and speculates on its potential for integrating control decisions in terms of IPPS.

Umar et al. (2015) proposed a hybrid GA for integrated scheduling, dispatching, and conflict-free routing of jobs and AGVs in a flexible manufacturing system environment, with the goal of optimizing makespan, AGV travel time, and penalty cost owing to tardiness, as well as conflict avoidance. Their algorithm uses an adaptive weight approach in the multi-objective fitness function and a fuzzy expert system to control the genetic operators.

Ramakurthi et al. (2021) proposed an integrated classifier-assisted evolutionary multi-objective evolutionary approach for improving IPPS in gear manufacturing industries in India, with the goals of minimizing makespan energy consumption and increasing service utilization rate, interoperability, and reliability. The approach involves using text-mining-based supervised ML models to classify suppliers, formulating the problem as a multi-objective MILP model, and optimizing the process planning and scheduling functions using a hybrid multi-objective moth flame optimization algorithm.

Zhai et al. (2021) proposed a holistic framework for integrating predictive maintenance models with production scheduling (PdM-IPS) to create added value and bridge the gap between research and industry. They suggested a conditional variational autoencoder-based generative deep learning model that can extract an operation-specific health indicator from large-scale industrial condition monitoring data. Using NASA's C-MAPSS dataset and real industrial data from machining centers they validated their model. The results show that it can detect and quantify changes in machine conditions to enable PdM-IPS.

11.8   CONCLUSION

In conclusion, this chapter overviewed solution techniques for integrated manufacturing functions. The techniques were divided into six groups: exact methods, heuristic methods, metaheuristic methods, agent-based methods, hybrid methods, and machine learning methods. Studies in the related literature were provided. Exact methods offer guaranteed optimal solutions but may not be feasible for large-scale or complex problems. Recent studies present advanced models that could be faster than traditional exact methods and better than metaheuristic solutions. Heuristic methods provide good solutions in a shorter time frame. Still, the solutions may not be optimal—metaheuristic methods, such as GA and SA, balance solution quality with computational time. Agent-based methods model the decision-making of individual entities within a system. Hybrid methods combine two or more solution techniques to take advantage of their strengths. ML methods use data-driven approaches to improve the performance of manufacturing systems. When selecting a technique for a specific integration problem for manufacturing functions, it is important to consider the trade-offs between solution quality and computational time. Whether using exact methods, metaheuristics, machine learning methods, or any other method, it is essential to follow the current developments in this area, as new approaches may provide better results in a shorter time.

REFERENCES

Amin-Naseri, M.R., Afshari, A.J., 2012. A hybrid genetic algorithm for integrated process planning and scheduling problem with precedence constraints. *The International Journal of Advanced Manufacturing Technology* 59, 273–287.

Assarzadegan, P., Rasti-Barzoki, M., 2016. Minimizing sum of the due date assignment costs, maximum tardiness and distribution costs in a supply chain scheduling problem. *Applied Soft Computing* 47, 343–356. https://doi.org/10.1016/j.asoc.2016.06.005

Ausaf, M.F., Gao, L., Li, X., Al Aqel, G., 2015. A priority-based heuristic algorithm (PBHA) for optimizing integrated process planning and scheduling problem. *Cogent Engineering* 2, 1070494.

Barzanji, R., Naderi, B., Begen, M.A., 2020. Decomposition algorithms for the integrated process planning and scheduling problem. *Omega* 93, 102025. https://doi.org/10.1016/j.omega.2019.01.003

Baykasoglu, A., Ozbakir, L., 2009. A grammatical optimization approach for integrated process planning and scheduling. *Journal of Intelligent Manufacturing* 20, 211–221. https://doi.org/10.1007/s10845-008-0223-0

Chan, F.T., Kumar, V., Tiwari, M.K., 2006. Optimizing the performance of an integrated process planning and scheduling problem: an AIS-FLC based approach, in: *2006 IEEE Conference on Cybernetics and Intelligent Systems*. IEEE, pp. 1–8.

Chryssolouris, G., Chan, S., Suh, N.P., 1985. An integrated approach to process planning and scheduling. *CIRP Annals* 34, 413–417.

Dorigo, M., Maniezzo, V., Colorni, A., 1996. Ant system: Optimization by a colony of cooperating agents. *IEEE Transactions on Systems, Man, and Cybernetics, Part B (Cybernetics)* 26, 29–41. https://doi.org/10.1109/3477.484436

Erden, C., Demir, H.I., Kökçam, A.H., 2019. Solving integrated process planning, dynamic scheduling, and due date assignment using metaheuristic algorithms. *Mathematical Problems in Engineering* 2019, 1–19. https://doi.org/10.1155/2019/1572614

Fu, Y., Hou, Y., Chen, Z., Pu, X., Gao, K., Sadollah, A., 2022. Modelling and scheduling integration of distributed production and distribution problems via black widow optimization. *Swarm and Evolutionary Computation* 68, 101015. https://doi.org/10.1016/j.swevo.2021.101015

Gao, S., Daaboul, J., Le Duigou, J., 2021. Process planning, scheduling, and layout optimization for multi-unit mass-customized products in sustainable reconfigurable manufacturing system. *Sustainability* 13, 13323. https://doi.org/10.3390/su132313323

Glover, F., 1986. Future paths for integer programming and links to artificial intelligence. *Computers & Operations Research, Applications of Integer Programming* 13, 533–549. https://doi.org/10.1016/0305-0548(86)90048-1

Goli, A., Davoodi, S.M.R., 2018. Coordination policy for production and delivery scheduling in the closed loop supply chain. *Production Engineering: Research and Development* 12, 621–631. https://doi.org/10.1007/s11740-018-0841-0

Gucdemir, H., Selim, H., 2017. Customer centric production planning and control in job shops: A simulation optimization approach. *Journal of Manufacturing Systems* 43, 100–116. https://doi.org/10.1016/j.jmsy.2017.02.004

Guo, Y.W., Li, W.D., Mileham, A.R., Owen, G.W., 2009. Applications of particle swarm optimisation in integrated process planning and scheduling. *Robotics and Computer-Integrated Manufacturing* 25, 280–288.

Holland, J.H., 1992. *Adaptation in Natural and Artificial Systems: An Introductory Analysis with Applications to Biology, Control, and Artificial Intelligence*. MIT Press, Cambridge, MA.

Huang, S.H., Zhang, H.-C., Smith, M.L., 1995. A progressive approach for the integration of process planning and scheduling. *IIE Transactions* 27, 456–464. https://doi.org/10.1080/07408179508936762

Kennedy, J., Eberhart, R., 1995. Particle swarm optimization, in: *Proceedings of ICNN'95 - International Conference on Neural Networks*, pp. 1942–1948, vol.4. https://doi.org/10.1109/ICNN.1995.488968

Kirkpatrick, S., Gelatt, C.D., Vecchi, M.P., 1983. Optimization by simulated annealing. *Science* 220, 671–680.

Lee, H., Kim, S.-S., 2001. Integration of process planning and scheduling using simulation based genetic algorithms. *The International Journal of Advanced Manufacturing Technology* 18, 586–590.

Leung, C.W., Wong, T.N., Mak, K.-L., Fung, R.Y., 2010. Integrated process planning and scheduling by an agent-based ant colony optimization. *Computers & Industrial Engineering* 59, 166–180.

Li, W., Han, D., Gao, L., Li, X., Li, Y., 2022. Integrated production and transportation scheduling method in hybrid flow shop. *Chinese Journal of Mechanical Engineering* 35, 12. https://doi.org/10.1186/s10033-022-00683-7

Li, W.D., McMahon, C.A., 2007. A simulated annealing-based optimization approach for integrated process planning and scheduling. *International Journal of Computer Integrated Manufacturing* 20, 80–95.

Li, X., Gao, L., 2020. A modified genetic algorithm based approach for IPPS, in: Li, X., Gao, L. (Eds.), *Effective Methods for Integrated Process Planning and Scheduling, Engineering Applications of Computational Methods*. Springer, Berlin, Heidelberg, pp. 209–233. https://doi.org/10.1007/978-3-662-55305-3_11

Li, X., Gao, L., Zhang, C., Shao, X., 2010. A review on integrated process planning and scheduling. *International Journal of Manufacturing Research* 5, 161–180.

Lihong, Q., Shengping, L., 2012. An improved genetic algorithm for integrated process planning and scheduling. *The International Journal of Advanced Manufacturing Technology* 58, 727–740. https://doi.org/10.1007/s00170-011-3409-0

Lin, C.-W.R., Chen, H.-Y., Xiao, Q.-S., 2001. Dynamic integrated process planning and scheduling. *Journal of the Chinese Institute of Industrial Engineers* 18, 21–32. https://doi.org/10.1080/10170660109509169

Liu, Q., Li, X., Gao, L., Fan, J., 2022a. A multi-MILP model collaborative optimization method for integrated process planning and scheduling problem. *IEEE Transactions on Engineering Management*. https://doi.org/10.1109/TEM.2022.3208431

Liu, Q., Li, X., Gao, L., Li, Y., 2021. A modified genetic algorithm with new encoding and decoding methods for integrated process planning and scheduling problem. *IEEE Transactions on Cybernetics* 51, 4429–4438. https://doi.org/10.1109/TCYB.2020.3026651

Liu, Q., Li, X., Gao, L., Wang, G., 2022b. A multiobjective memetic algorithm for integrated process planning and scheduling problem in distributed heterogeneous manufacturing systems. *Memetic Computing* 14, 193–209. https://doi.org/10.1007/s12293-022-00364-x

Mahmud, S., Chakrabortty, R.K., Abbasi, A., Ryan, M.J., 2022. Swarm intelligent based metaheuristics for a bi-objective flexible job shop integrated supply chain scheduling problems. *Applied Soft Computing* 121, 108794. https://doi.org/10.1016/j.asoc.2022.108794

Maravelias, C.T., Sung, C., 2009. Integration of production planning and scheduling: Overview, challenges and opportunities, in: *Computers & Chemical Engineering, FOCAPO 2008 – Selected Papers from the Fifth International Conference on Foundations of Computer-Aided Process Operations 33*, pp. 1919–1930. https://doi.org/10.1016/j.compchemeng.2009.06.007

Nejad, H.T.N., Sugimura, N., Iwamura, K., 2011. Agent-based dynamic integrated process planning and scheduling in flexible manufacturing systems. *International Journal of Production Research* 49, 1373–1389. https://doi.org/10.1080/00207543.2010.518741

Nejad, H.T.N., Sugimura, N., Iwamura, K., Tanimizu, Y., 2008. Integrated dynamic process planning and scheduling in flexible manufacturing systems via autonomous agents. *JAMDSM* 2, 719–734. https://doi.org/10.1299/jamdsm.2.719

Park, S., Gervasio, M., Shaw, M., Dejong, G., 1993. Explanation-based learning for intelligent process planning. *IEEE Transactions on Systems, Man, and Cybernetics* 23, 1597–1616. https://doi.org/10.1109/21.257757

Pereira, M.T., Nagano, M.S., 2022. Hybrid metaheuristics for the integrated and detailed scheduling of production and delivery operations in no-wait flow shop systems. *Computers & Industrial Engineering* 170, 108255. https://doi.org/10.1016/j.cie.2022.108255

Phanden, R.K., Jain, A., Davim, J.P., 2019. *Integration of Process Planning and Scheduling: Approaches and Algorithms*. Boca Raton, FL, CRC Press.

Ramakurthi, V.B., Manupati, V.K., Machado, J., Varela, L., 2021. A hybrid multi-objective evolutionary algorithm-based semantic foundation for sustainable distributed manufacturing systems. *Applied Sciences* 11, 6314. https://doi.org/10.3390/app11146314

Rossi, A., Lanzetta, M., 2020. Integration of hybrid additive/subtractive manufacturing planning and scheduling by metaheuristics. *Computers & Industrial Engineering* 144, 106428. https://doi.org/10.1016/j.cie.2020.106428

Rostami, M., Nikravesh, S., Shahin, M., 2020. Minimizing total weighted completion and batch delivery times with machine deterioration and learning effect: A case study from wax production. *Operations Research* 20, 1255–1287. https://doi.org/10.1007/s12351-018-0373-6

Shafiee-Gol, S., Kia, R., Tavakkoli-Moghaddam, R., Kazemi, M., Kamran, M.A., 2021. Integration of parts scheduling, MRP, production planning and generalized fixed-charge transportation planning in the design of a dynamic cellular manufacturing system. *RAIRO – Operations Research* 55, S1875–S1912. https://doi.org/10.1051/ro/2020062

Sharifi, M., Taghipour, S., 2021. Optimal production and maintenance scheduling for a degrading multi-failure modes single-machine production environment. *Applied Soft Computing* 106, 107312. https://doi.org/10.1016/j.asoc.2021.107312

Shi, H., Chu, Y., You, F., 2014. Integrated planning, scheduling, and dynamic optimization for continuous processes, in: *53rd IEEE Conference on Decision and Control*. IEEE, Los Angeles, CA, USA, pp. 388–393. https://doi.org/10.1109/CDC.2014.7039412

Shukla, S.K., Tiwari, M.K., Son, Y.J., 2007. Bidding-based multi-agent system for integrated process planning and scheduling: A data-mining and hybrid tabu-SA algorithm-oriented approach. *The International Journal of Advanced Manufacturing Technology* 38, 163. https://doi.org/10.1007/s00170-007-1087-8

Umar, U.A., Ariffin, M.K.A., Ismail, N., Tang, S.H., 2015. Hybrid multiobjective genetic algorithms for integrated dynamic scheduling and routing of jobs and automated-guided vehicle (AGV) in flexible manufacturing systems (FMS) environment. *The International Journal of Advanced Manufacturing Technology* 81, 2123–2141. https://doi.org/10.1007/s00170-015-7329-2

Wang, J., Yao, S., Sheng, J., Yang, H., 2019. Minimizing total carbon emissions in an integrated machine scheduling and vehicle routing problem. *Journal of Cleaner Production* 229, 1004–1017. https://doi.org/10.1016/j.jclepro.2019.04.344

Wang, J.-B., Guo, Q., 2010. A due-date assignment problem with learning effect and deteriorating jobs. *Applied Mathematical Modelling* 34, 309–313. https://doi.org/10.1016/j.apm.2009.04.020

Wong, T.N., Leung, C.W., Mak, K.L., Fung, R.Y.K., 2006. Integrated process planning and scheduling/rescheduling—An agent-based approach. *International Journal of Production Research* 44, 3627–3655. https://doi.org/10.1080/00207540600675801

Xia, H., Li, X., Gao, L., 2016. A hybrid genetic algorithm with variable neighborhood search for dynamic integrated process planning and scheduling. *Computers & Industrial Engineering* 102. https://doi.org/10.1016/j.cie.2016.10.015

Yu, M.R., Yang, B., Chen, Y., 2018. Dynamic integration of process planning and scheduling using a discrete particle swarm optimization algorithm. *Advances in Production Engineering & Management* 13, 279–296. https://doi.org/10.14743/apem2018.3.290

Zhai, S., Gehring, B., Reinhart, G., 2021. Enabling predictive maintenance integrated production scheduling by operation-specific health prognostics with generative deep learning. *Journal of Manufacturing Systems* 61, 830–855. https://doi.org/10.1016/j.jmsy.2021.02.006

Zhang, L., Wong, T.N., Fung, R.Y.K., 2012. A multi-agent system for dynamic integrated process planning and scheduling using heuristics, in: Jezic, G., Kusek, M., Nguyen, N.-T., Howlett, R.J., Jain, L.C. (Eds.), *Agent and Multi-Agent Systems. Technologies and Applications*, Lecture Notes in Computer Science. Springer, Berlin, Heidelberg, pp. 309–318. https://doi.org/10.1007/978-3-642-30947-2_35

Zhang, X., Zhang, H., Yao, J., 2020. Multi-objective optimization of integrated process planning and scheduling considering energy savings. *Energies* 13, 6181. https://doi.org/10.3390/en13236181

Zhang, Z., Benyoucef, L., Sialat, A., 2022. Sustainable integrated process planning and scheduling (IPPS) in RMS: Past, present and future. *IFAC-PapersOnLine* 55, 791–797. https://doi.org/10.1016/j.ifacol.2022.09.506

Zhao, F., Zhu, A., Yu, D., Yang, Y., 2006. A hybrid particle swarm optimization (PSO) algorithm schemes for integrated process planning and production scheduling, in: *2006 6th World Congress on Intelligent Control and Automation*, pp. 6772–6776. https://doi.org/10.1109/WCICA.2006.1714395

12 Integrated Process Planning, Scheduling, Due-Date Assignment, and Delivery

12.1 INTRODUCTION

Process planning, scheduling, and due-date assignment functions were mentioned earlier in the book, but another critical function, delivery, will be discussed in this chapter. The more each function is unaware of other functions and is solved on its own, the more these functions try to optimize local benefits. Because they do not consider the global benefit, the global performance will be adversely affected. In order to optimize global performance and take care of global benefits, each sub-function should be integrated and optimized together. Since each function creates an input for other downstream functions, bad inputs will come to sub-functions if they are not optimized together. Individually handled functions and local optimum will worsen the performance of other functions, and overall performance will deteriorate significantly.

In the previous chapters, we mentioned that some functions are integrated in pairs. Examples of pairwise integrations are the IPPS (Integrated Process Planning and Scheduling) (Barzanji et al., 2020), the SWDDA (Scheduling with Due-date Assignment) (Zhao et al., 2018), and the integrated production and distribution scheduling (IPDS) problems (Fu et al., 2017; Karaoğlan and Kesen, 2017). Again, these chapters mentioned that integrating process planning, scheduling, and due-date assignment functions is a relatively new subject. This problem has been addressed in only a few studies in IPPSDDA (Integrated Process Planning and Scheduling and Due-date Assignment) (Demir et al., 2015; Demir and Erden, 2017; Demir et al., 2021), and DIPPSDDA (Dynamic IPPSDDA) problems (Erden et al., 2019, 2021; Demir and Erden, 2020). In this chapter, to the best of our knowledge, it will be mentioned that integrating the four functions is an entirely new topic and a subject suitable for many studies. The Integrated Process Planning, Scheduling, Due-date Assignment, and Delivery (IPPSDDAD) problem is the highest integration level that can be solved among the integrations.

12.1.1 DELIVERY

The issue that needs to be addressed after production planning is delivery. The delivery of manufactured products to customers is a significant supply chain step.

DOI: 10.1201/9781003215295-12

Although delivery is a kind of vehicle routing problem (VRP), in many SWDDA or SWDWA (Scheduling with Due-window Assignment) problems in the literature, delivery only refers to when the products are sent from the company in batches or are ready for delivery. In these studies, it is assumed that the delivery in batches has a fixed cost or varies according to the number of products in the batch (Yin et al., 2014). These assumptions are made for modeling simplicity and are not entirely correct. Delivery costs also vary depending on the distance of the customers from the company and the distance from each other. These costs may also vary depending on peak traffic hours and highway tolls. In the U.S., annual spending on non-military logistics accounts for more than 11% of the Gross National Product. In addition, for many products, more than 30% of the cost of goods sold is logistics expenses, and transportation costs constitute a significant percentage of logistics (Hall and Potts, 2003).

Let's consider the IPDS problems after the SWDDA and SWDWA problems. The delivery part may include the part until the delivery of the products to the customers. Sometimes, it covers the part until it is ready for delivery and assumes that the 3PL (Third Party Logistics) companies make the delivery as outsourcing. In this case, the delivery part other than single item delivery and single customer delivery becomes a type of VRP problem. In the VRP problem, delivery and pickup operations can be mixed or separated. In these problems, some objective functions are tried to be minimized such as distance covered, the total delivery time or transportation cost are minimized. Sometimes, carbon dioxide emissions are minimized, and green supply chain objectives are tried to be achieved (Fu et al., 2017; Karaoğlan and Kesen, 2017).

12.1.2 VEHICLE ROUTING

In SWDDA and SWDWA problems, we have said that delivery refers to the situation until the products are ready for delivery in batches or until they start delivery. But in fact, delivery also includes the delivery of products to customers. In this case, we consider delivery as a VRP problem except for single-item and single-customer delivery. In the classical VRP problem, the following assumptions can be made: the vehicles start delivery from a center (depot), visit the customers, and return to the firm (depot). At delivery, each customer is assigned to only one vehicle, and vehicles have capacity constraints.

Here, special cases of the VRP problem occur according to each assumption. Sometimes the VRP problem with one vehicle and no capacity constraint is solved. In other cases, the VRP problem with a capacity constraint and delivery made by multi-trip is solved, and this problem is known as Multi-Trip VRP (MTVRP). In some cases, there is a fleet of vehicles; in this case, the vehicles can be homogeneous or heterogeneous. Furthermore, time windows can be considered; these problems are known as VRP with time windows (VRPTW). The time window can be hard, soft, or sometimes mixed in such problems. Time window restrictions can vary based on customers, roads, depots (hubs), vehicles, and drivers (Eksioglu et al., 2009; Braekers et al., 2016; Dixit et al., 2019).

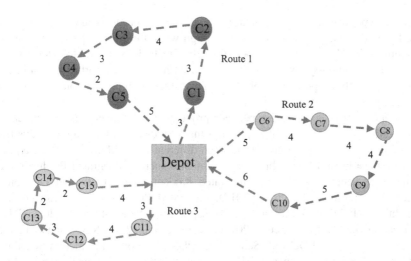

FIGURE 12.1 Simple VRP problem and a solution.

VRP problems are also handled based on backhaul status. Such problems are called VRP with Backhaul (VRPB) problems. In some problems, there may be simultaneous pick-up and delivery, while in some cases, there may be only linehaul or backhaul. Suppose simultaneous pick-up and delivery are to be made. In that case, these problems are called VRP with simultaneous Pick-up and Delivery (VRPPD) problems (Eksioglu et al., 2009; Braekers et al., 2016; Dixit et al., 2019).

If the vehicle does not return to the depot after delivery, these problems are called Open VRP (OVRP). The problems are called Dynamic VRP (DVRP) problems if customer demands can occur dynamically during transportation. If the problem has more than one depot, then the problem is called a Multi-Depot VRP (MDVRP) problem. Some problems have capacity constraints on vehicles, called capacitated VRP (CVRP) problems. A simple VRP problem is shown in Figure 12.1.

In addition, in VRP problems, it is sometimes obvious how many customers will be visited; some are known in advance, but some can be probabilistically changed. In some problems, demands can be split or delivered as a whole without splitting. Sometimes the customer demand quantity is deterministic, and sometimes, the demand quantity is stochastic.

12.1.3 PRODUCTION AND DISTRIBUTION SCHEDULING

The intense competition in the world has forced companies to be more coordinated in the production and delivery stages. Although make-to-stock (MTS) production was common in the past, the spread of the JIT (Just in Time) philosophy has led to the spread of make-to-order (MTO) production. For example, if white goods production in China is considered, while the order response time used to be 30 days before, this time has been reduced to 7 days with the widespread use of make-to-order production (Zou et al., 2018).

MTO-type production is mandatory for perishable products and some products. The finished product inventory was very high when the MTS-type production was common. After the MTO-type production, the finished good inventory level started to remain substantially low. The order response time must be short for low inventory levels, especially for perishable products. This situation imposes strong coordination between production and delivery.

The necessity of coordination between production and delivery and the increase in this coordination have led to the spread of many studies in the literature in this field, with almost the same meaning but between different titles. Many integrated types of research have been done in this area with names such as Integrated Production and Distribution Scheduling (IPDS), Integrated Production and Outbound Distribution Scheduling (IPODS) (Fu et al., 2017), Integrated Production and Transportation Scheduling (IPTS), Production and Transportation Scheduling Problem (PTSP) (Geismar et al., 2008; Reiter et al., 2011; Lacomme et al., 2016; Karaoğlan and Kesen, 2017), Production and Delivery Scheduling (PDS), Integrated Production Scheduling and Vehicle Routing Problem (IPSVRP) (Zou et al., 2018), and Production and Delivery Scheduling Problem with Time Windows (PDPTW) (Garcia and Lozano, 2005).

Some examples of perishable products or products with a short life span are food and ready-mix concrete. In addition, in systems where the direct order system is an important part of the business where customization is very high, for example, in the computer and food catering service industry and MTO production systems, the stock is not kept or kept at minimum levels and production and delivery coordination is very important (Chen and Vairaktarakis, 2005).

Batch and lot sizing are important issues in systems where production and delivery are integrated. Optimum batch and lot sizing are done for the following reasons: (a) for minimization of production setup, (b) if products are to be sent to the same customer in a batch, (c) to minimize the transportation costs, (d) and to produce the same batch products simultaneously. For these reasons, batch and lot sizing determination are critical (Hall and Potts, 2005).

In addition, the sub-topics addressed in these problems may include time intervals to be followed in delivery, time intervals in production, machines may not be available at specific times, fixed delivery dates, committed delivery dates, capacity constraints on delivery vehicles, and vehicles can form a homogeneous or heterogeneous fleet.

Finally, in the studies on these subjects, sometimes the production planning and the part until the batches are ready for shipment, and sometimes the part until the moment of loading to the vehicle is handled in an integrated manner. In some studies, together with production planning, the part of the delivery is up to the customer. That is, the vehicle routing part is solved concurrently.

12.2 INTEGRATED PROCESS PLANNING, SCHEDULING, DUE-DATE ASSIGNMENT, AND DELIVERY

When the literature is examined, many studies are on SWDDA, IPPS, and IPDS. Although not very common, studies have also been done on the IPPSDDA problem. However, the four crucial functions' integration, which are the process planning,

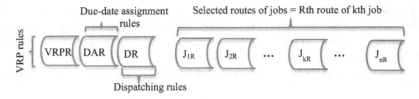

FIGURE 12.2 Chromosome representation of IPPSDDAD problem.

scheduling, due-date assignment, and delivery functions, was not done. Higher-level integrations will be beneficial for increasing global performance.

Integration is critical as these four functions influence each other as in chain rings, and the upstream functions provide inputs to the downstream functions. If each function is optimized individually, each will struggle for its local optimization, and downstream functions will receive bad input, unaware of global performance. Therefore, functions that are optimized individually and follow local optimization will cause inferior global performance.

Since only the scheduling and VRP problem are alone NP-Hard problems, the problem of solving the four functions together by integrating the process planning and due-date determination functions will naturally be an NP-Hard problem. For this reason, metaheuristic approaches will be appropriate for the upper integration we mentioned, as in the sub-integration above. As in IPPSDDA problem-solving, the chromosome structure in Figure 12.2 above can be used in metaheuristic approaches in IPPSDDAD problem-solving. As VRP heuristic rules, savings algorithms, the sweep algorithm, the nearest neighbor algorithm, and others can be used for the first gene in the chromosome (Laporte et al., 2000).

12.3 LITERATURE SURVEY

In the IPPSDDAD problem, four functions will be integrated. To the best of our knowledge, this level of integration is a topic that has not been addressed in the literature. There are many studies on pairwise integrations, such as SWDDA, IPPS, and IPDS, in the literature, but not all of them will be mentioned here. In particular, IPPS, SWDDA, and SWDWA topics, which have been extensively covered in previous chapters, will not be covered here. In this chapter, the research is directed at the delivery. IPDS and VRP problems that were not mentioned in the earlier chapters will be briefly discussed here.

For research on the SWDDA, Chapter 7 can be reviewed. Chapter 8 can be visited for the SWDWA problem. A broad description of the IPPS studies is given in Chapter 5. A detailed explanation and literature review are given in Chapter 9 on the IPPSDDA studies and Chapter 10 on the DIPPSDDA studies. In this chapter, before discussing the IPDS and the VRP subjects, which were not mentioned in the previous chapters, it will be helpful to give some of the studies on the IPPSDDA (Demir and Erden, 2017; Demir et al., 2021) literature on the DIPPSDDA problem (Erden et al., 2019, 2021; Demir and Erden, 2020).

Delivery problem is a group of problems closely related to the problem called VRP. Altinkemer and Gavish (1991) used parallel savings-based heuristics to solve the delivery problem. Many studies have integrated the delivery problem with the scheduling problem. Liu and Lu (2016) addressed the scheduling problem that simultaneously considers production and job delivery with machine availability constraints.

The studies of Dantzig et al. (1954) and Dantzig and Ramser (1959) can be given as the first studies on VRP. As a first study, G. Dantzig et al. (1954) studied the solution to a large-scale traveling-salesman problem. In the following years, G. B. Dantzig and Ramser (1959) studied the truck dispatching problem. These studies are among the earliest studies on the VRP problem. Then, new variations of the VRP problem were studied. For example, Solomon (1984, 1987) studied the VRP problem with time window constraints in different years. Before going into detail about VRP, it will be beneficial to examine the taxonomic review study on VRP by Eksioglu et al. (2009).

There are numerous studies and many surveys on VRP. It will be handy to examine Braekers et al. (2016), Mor and Speranza (2022) survey studies, and Golden et al. (2008), Toth and Vigo (2014) books, for more detailed information on this subject. Besides, for more information on dynamic and stochastic VRP, it is better to see Ritzinger et al. (2016). For VRP with multiple depots Montoya-Torres et al. (2015) can be examined. For VRP with time windows problem, readers may refer to Dixit et al. (2019). For simultaneous pickup and delivery VRP problems, readers may refer to Montané and Galvão (2006).

Although there is no study on IPPSDDAD, the pairwise integrations of functions have been intensively studied. There are also many studies on scheduling and delivery. Many studies have investigated this integration as an IPTS or PTSP problem (Reiter et al., 2011; Lacomme et al., 2016). In many studies, this problem has been defined as PDS, IPDS, or IPODS problem (Chen and Hall, 2022). Some studies have studied the problem as an IPSVRP problem. As PDPTW, the problem with time window constraints is also investigated (Garcia and Lozano, 2005).

As there are perishable products in daily life, and as a result of the spread of JIT philosophy in production and the spread of MTO-type production instead of MTS-type production, the necessity of coordination between production scheduling and distribution is constantly increasing. As a result of either not keeping stock of finished goods or keeping this stock at minimum levels, production and distribution scheduling coordination is much more critical than ever. This coordination also applies to customer satisfaction, and this coordination and integration is a must in the competitive world.

There are studies on the integration of production planning and delivery for products that are perishable and have a short useful life (Geismar et al., 2008; Ma et al. (2019). In some studies, only the time of the completion of the production of the products in the delivery part or the part until the time of loading into the vehicle and starting the delivery has been tried to be integrated and optimized (Leung and Chen, 2013). Some studies integrate delivery with production planning, including vehicle routing (Lacomme et al., 2016).

12.4 CONCLUSION

Process planning, scheduling, due-date assignment, and delivery are four critical manufacturing and supply chain functions, in which their integration is studied in the literature. These functions and their pairwise and triple integration are discussed in detail in this book. SWDDA, SWDWA, IPPS, PTSP, IPODS, IPPSDDA, and DIPPSDDA problems are explained, and studies on these integrations are mentioned in detail.

As far as we know, the IPPSDDAD problem has not been addressed, and these four functions have not been considered as a whole in the literature. Since these four functions have important effects on each other, and the solutions and results in the upstream will be inputs to the downstream, we need to consider the following functions while solving these functions. These functions may have conflicting goals and affect each other adversely. That's why, instead of solving these four functions sequentially, it would be more logical to consider these functions in an integrated way and to look for an integrated solution focused only on global optimization instead of local optimizations.

Since only the scheduling and vehicle routing parts are stand-alone NP-Hard problems, the integrated handling of the four functions would be much more complex and naturally be an NP-Hard problem. Since exact solutions are not possible except for minimal problems, it would be appropriate to use metaheuristic methods in quadruple integration as in other pairwise and triple integrations.

In many VRP problems and IPODS or PTSP problems, the due dates or the time windows are not optimized together with the problem. This problem is tried to be solved according to the customer's due dates or time windows. However, in SWDDA and SWDWA problems, the due dates or due windows are decided as a result of optimization. In this case, the customer does not determine the delivery dates or windows. Either the company gives itself the most suitable delivery dates, or the customer and the company try to determine the most suitable dates for both parties. In the proposed IPPSDDAD or IPPSDWAD (Integrated Process Planning, Scheduling, Due-window Assignment, and Delivery) problems, the most appropriate delivery dates or delivery windows can be determined for the company or both parties. It can be optimized together with the problem. Customer weights can also be included in the problem, and important customers can be given closer due dates or due windows. These customers can be given priority in production scheduling and delivery. Thus, significant gains can be achieved in terms of weighted performance.

REFERENCES

Altinkemer, K., & Gavish, B. (1991). Parallel savings based heuristics for the delivery problem. *Operations Research*, *39*(3), 456–469. https://doi.org/10.1287/opre.39.3.456

Barzanji, R., Naderi, B., & Begen, M. A. (2020). Decomposition algorithms for the integrated process planning and scheduling problem. *Omega*, *93*, 102025. https://doi.org/10.1016/j.omega.2019.01.003

Braekers, K., Ramaekers, K., & Van Nieuwenhuyse, I. (2016). The vehicle routing problem: State of the art classification and review. *Computers & Industrial Engineering*, *99*, 300–313. https://doi.org/10.1016/j.cie.2015.12.007

Chen, Z. L., & Hall, N. G. (2022). Integrated production and outbound distribution scheduling: Offline problems. In *Supply Chain Scheduling* (pp. 53–136). Springer, Cham. https://doi.org/10.1007/978-3-030-90374-9_3

Chen, Z.-L., & Vairaktarakis, G. L. (2005). Integrated scheduling of production and distribution operations. *Management Science*, *51*(4), 614–628. https://doi.org/10.1287/mnsc.1040.0325

Dantzig, G. B., & Ramser, J. H. (1959). The truck dispatching problem. *Management Science*, *6*(1), 80–91.

Dantzig, G., Fulkerson, R., & Johnson, S. (1954). Solution of a large-scale traveling-salesman problem. *Journal of the Operations Research Society of America*, *2*(4), 393–410.

Demir, H. I., Cakar, T., Ipek, M., Uygun, O., & Sari, M. (2015). Process planning and due-date assignment with ATC dispatching where earliness, tardiness and due-dates are punished. *Journal of Industrial and Intelligent Information*, *3*(3), 197–204. https://doi.org/10.12720/jiii.3.3.197-204

Demir, H. I., & Erden, C. (2017). Solving process planning and weighted scheduling with WNOPPT weighted due-date assignment problem using some pure and hybrid meta-heuristics. *SAU Journal of Science*, *21*(2), 1–1. https://doi.org/10.16984/saufenbilder.297014

Demir, H. I., & Erden, C. (2020). Dynamic integrated process planning, scheduling and due-date assignment using ant colony optimization. *Computers & Industrial Engineering*, *149*, 106799. https://doi.org/10.1016/j.cie.2020.106799

Demir, H. I., Phanden, R., Kökçam, A., Erkayman, B., & Erden, C. (2021). Hybrid evolutionary strategy and simulated annealing algorithms for integrated process planning, scheduling and due-date assignment problem. *Academic Platform Journal of Engineering and Science*, *9*(1), 86–91. https://doi.org/10.21541/apjes.764150

Dixit, A., Mishra, A., & Shukla, A. (2019). Vehicle routing problem with time windows using meta-heuristic algorithms: A survey. In N. Yadav, A. Yadav, J. C. Bansal, K. Deep, & J. H. Kim (Eds.), *Harmony Search and Nature Inspired Optimization Algorithms* (Vol. 741, pp. 539–546). Springer, Singapore. https://doi.org/10.1007/978-981-13-0761-4_52

Eksioglu, B., Vural, A. V., & Reisman, A. (2009). The vehicle routing problem: A taxonomic review. *Computers & Industrial Engineering*, *57*(4), 1472–1483. https://doi.org/10.1016/j.cie.2009.05.009

Erden, C., Demir, H. I., & Canpolat, O. (2021). A modified integer and categorical PSO algorithm for solving integrated process planning, dynamic scheduling and due date assignment problem. *Scientia Iranica*. https://doi.org/10.24200/sci.2021.55250.4130

Erden, C., Demir, H. I., & Kökçam, A. H. (2019). Solving integrated process planning, dynamic scheduling, and due date assignment using metaheuristic algorithms. *Mathematical Problems in Engineering*, *2019*, 1–19. https://doi.org/10.1155/2019/1572614

Fu, L.-L., Aloulou, M. A., & Triki, C. (2017). Integrated production scheduling and vehicle routing problem with job splitting and delivery time windows. *International Journal of Production Research*, *55*(20), 5942–5957. https://doi.org/10.1080/00207543.2017.1308572

Garcia, J. M., & Lozano, S. (2005). Production and delivery scheduling problem with time windows. *Computers & Industrial Engineering*, *48*(4), 733–742. https://doi.org/10.1016/j.cie.2004.12.004

Geismar, H. N., Laporte, G., Lei, L., & Sriskandarajah, C. (2008). The integrated production and transportation scheduling problem for a product with a short lifespan. *INFORMS Journal on Computing*, *20*(1), 21–33.

Golden, B., Raghavan, S., & Wasil, E. (Eds.) (2008). *The Vehicle Routing Problem: Latest Advances and New Challenges* (Vol. 43). Springer US. https://doi.org/10.1007/978-0-387-77778-8

Hall, N. G., & Potts, C. N. (2003). Supply chain scheduling: Batching and delivery. *Operations Research*, *51*(4), 566–584. https://doi.org/10.1287/opre.51.4.566.16106

Hall, N. G., & Potts, C. N. (2005). The coordination of scheduling and batch deliveries. *Annals of Operations Research*, *135*(1), 41–64. https://doi.org/10.1007/s10479-005-6234-8

Karaoğlan, I ., & Kesen, S. E. (2017). The coordinated production and transportation scheduling problem with a time-sensitive product: A branch-and-cut algorithm. *International Journal of Production Research*, *55*(2), 536–557. https://doi.org/10.1080/00207543.2016.1213916

Lacomme, P., Moukrim, A., Quilliot, A., & Vinot, M. (2016). The integrated production and transportation scheduling problem based on a GRASP×ELS resolution scheme. *IFAC-PapersOnLine*, *49*(12), 1466–1471. https://doi.org/10.1016/j.ifacol.2016.07.778

Laporte, G., Gendreau, M., Potvin, J.-Y., & Semet, F. (2000). Classical and modern heuristics for the vehicle routing problem. *International Transactions in Operational Research*, *7*(4–5), 285–300. https://doi.org/10.1111/j.1475-3995.2000.tb00200.x

Leung, J. Y.-T., & Chen, Z.-L. (2013). Integrated production and distribution with fixed delivery departure dates. *Operations Research Letters*, *41*(3), 290–293. https://doi.org/10.1016/j.orl.2013.02.006

Liu, P., & Lu, X. (2016). Integrated production and job delivery scheduling with an availability constraint. *International Journal of Production Economics*, *176*, 1–6. https://doi.org/10.1016/j.ijpe.2016.03.006

Montané, F. A. T., & Galvão, R. D. (2006). A tabu search algorithm for the vehicle routing problem with simultaneous pick-up and delivery service. *Computers & Operations Research* 33(3), 595–619.

Montoya-Torres, J. R., López Franco, J., Nieto Isaza, S., Felizzola Jiménez, H., & Herazo-Padilla, N. (2015). A literature review on the vehicle routing problem with multiple depots. *Computers & Industrial Engineering*, *79*, 115–129. https://doi.org/10.1016/j.cie.2014.10.029

Mor, A., & Speranza, M. G. (2022). Vehicle routing problems over time: A survey. *Annals of Operations Research*, *314*(1), 255–275. https://doi.org/10.1007/s10479-021-04488-0

Reiter, B. S., Makuschewitz, T., Novaes, A. G. N., Frazzon, E. M., & Lima, O. F. Jr. (2011). An approach for the sustainable integration of production and transportation scheduling. *International Journal of Logistics Systems and Management*, *10*(2), 158–179. https://doi.org/10.1504/IJLSM.2011.042626

Ritzinger, U., Puchinger, J., & Hartl, R. F. (2016). A survey on dynamic and stochastic vehicle routing problems. *International Journal of Production Research*, *54*(1), 215–231. https://doi.org/10.1080/00207543.2015.1043403

Solomon, M. M. (1984). *Vehicle Routing and Scheduling with Time Window Constraints: Models and Algorithms*. (No. 84-17364 UMI). (Doctoral dissertation, University of Pennsylvania).

Solomon, M. M. (1987). Algorithms for the vehicle routing and scheduling problems with time window constraints. *Operations Research*, *35*(2), 254–265. https://doi.org/10.1287/opre.35.2.254

Toth, P., & Vigo, D. (Eds.) (2014). *Vehicle Routing: Problems, Methods, and Applications* (2nd edition). Society for Industrial and Applied Mathematics: Mathematical Optimization Society, Philadelphia.

Yin, Y., Cheng, T. C. E., Wu, C.-C., & Cheng, S.-R. (2014). Single-machine batch delivery scheduling and common due-date assignment with a rate-modifying activity. *International Journal of Production Research*, *52*(19), 5583–5596. https://doi.org/10.1080/00207543.2014.886027

Zhao, C., Hsu, C.-J., Lin, W.-C., Liu, S.-C., & Yu, P.-W. (2018). Due date assignment and scheduling with time and positional dependent effects. *Journal of Information and Optimization Sciences*, *39*(8), 1613–1626. https://doi.org/10.1080/02522667.2017.1367515

Zou, X., Liu, L., Li, K., & Li, W. (2018). A coordinated algorithm for integrated production scheduling and vehicle routing problem. *International Journal of Production Research*, *56*(15), 5005–5024. https://doi.org/10.1080/00207543.2017.1378955

Index

Pages in *italics* refer figures and pages in **bold** refer tables.

Printed in the United States
by Baker & Taylor Publisher Services